How to Talk to Robots

How to Talk to Robots

A Girls' Guide to a Future
Dominated by AI

Tabitha Goldstaub

4th Estate • London

4th Estate
An imprint of HarperCollins*Publishers*
1 London Bridge Street
London SE1 9GF

www.4thEstate.co.uk

First published in Great Britain in 2020 by 4th Estate

1

A catalogue record for this book is available from the British Library

ISBN 978-0-00-840587-8 (hardback)
ISBN 978-0-00-832820-7 (trade paperback)

Printed and bound by CPI Group (UK) Ltd, Croydon

MIX
Paper from
responsible sources
FSC
www.fsc.org
FSC™ C007454

This book is produced from independently certified FSC paper
to ensure responsible forest management

Find out more about HarperCollins and the environment at
www.harpercollins.co.uk/green

For all the women I've met on the way

CONTENTS

CONTENTS

Me, Tabitha Goldstaub

FOREWORD

Growing up, my brother and I played Gameboy incessantly.
Yoshi's Dream World, *Mario Kart Racers*, *Tekken*. The gaming
world was full of beautiful pixelated landscapes. I loved the
thrill of the win. For me, gaming was a non-gendered space.
My household was one dedicated to fashion: my mother was
a magazine editor, my father a textiles merchant. So I grew up
sitting on the floor of fashion shows, but often more entranced
by my Gameboy than the clothes on the catwalk.

In 1995, when I was ten, my parents bought me one of the very
early home computers. It was proudly positioned in the kitchen,
where I would spend hours absorbed in the Microsoft Paint
program, creating personalised desktop backgrounds for each
member of my family. I felt like a tech Picasso – computers were
for art not engineering.

At school, the door to the Internet was Ask Jeeves, a British butler
who fronted the search engine, and my guide to Word and Excel
was Clippy, Microsoft's helpful paperclip mascot. I remember
having a Tamagotchi that was neither boy nor girl. All this is to
say that I was happily unaware of the notion that gender could

ever play a part in technology. It was not a male or female thing: it was *my* thing.

Cut to a few years later, and during mandatory GCSE IT classes, I was bored out of my brain. We were being shown how to run a doctor's reception and were using mail merge to access patients' data and send emails. I was not inspired. Thinking about it now, it's little wonder I was unable to see technology's true potential. I knew it could make a job as a receptionist more efficient and less repetitive, but what if my teachers had explained to me that tech could help predict which of my patients would need life-saving treatment? Maybe I would have gone on to study computer science. Instead, I followed in the footsteps of my grandpa and set about joining the advertising world. I was genuinely fascinated by the clever people behind the slogans, billboards and TV ads that captivated mine and my friends' impressionable minds. At that point I was enamoured and not yet fearful of their power.

Before university, I was working in a pub when brand new fully computerised till systems were installed. The manager decided to train the men on the tills first, which meant they became better faster and so were assigned more shifts behind the bar. The women ended up waiting tables. I didn't even stop to think about whether this division was fair – I thought computer stuff wasn't for me, and I think I just accepted that the guys should do the tills. But the consequence of this mindset meant I inadvertently accepted lower pay: waitresses earned less than bar staff, and I ended up hating a job I had come to love. What on earth had happened to me in those interim teenage years?! I still don't know at what point it became the norm that the world of computers was a male one.

By the time I arrived at university to study advertising, new developments with the Internet had set the world alight. Ask Jeeves was no longer the gatekeeper to information – now I could research anything and share it immediately. Users had also started communicating with each other directly, and I remember waiting for my university email to grant me access to The Facebook (as it was known back then), which was in its infancy as an American college social-networking platform. The nature of communication was changing, and my chosen route had begun to feel outdated. People's minds were not going to be made up by old-school advertising. I was falling in love with the Internet, with the endless opportunities it presented and its immediacy and scale.

Over the course of those three years, I chose modules that would enable me to redefine my relationship with computers. This included how to code. But I found that my dyslexia and my lack of confidence held me back. I was, quite honestly, hopeless. My real 'aha' moment came when I was introduced to Squarespace, a tool that meant I didn't have to write code in order to create a webpage and reach people through the Internet. There were companies and products being released every day that meant people like me, who didn't get on with coding, could still benefit hugely from opportunities available on the web.

All the advertising agencies at which I interned during my degree treated digital campaigns as the second division, and often relegated women to the same position. I felt pretty downtrodden. But the start-up world seemed to have a different vibe. From afar, I'd watched Martha Lane Fox, the co-founder of lastminute.com (who you'll read about later), take equal billing with her male co-founder as they reached millions online. It seemed like a level playing field, and this excited me endlessly.

So, in 2008, on my hunt to either start a business or work for a start-up, I met Charlie Muirhead, the CEO and Founder of t5m, a new style of digital studio that co-produced awesome online video content with celebrities. I joined as an intern, learning how to add words and codes to propel each video to the top ranks of YouTube and then distribute them across the Internet. When I was promoted to managing a team of other recent graduates, we soon realised that the more laborious, mechanical aspects of our job had to become automated for full efficiency – and to avoid us going mad! I relished the opportunity to work more closely with the engineering teams to make this happen.

During this time, the production part of the business had some great successes, but our part-human part-software team were truly flying. The business pivoted, and we concentrated solely on building the technology to enable us to distribute other people's videos. Seeing those online views rack up was everything to me. I was hooked: how could we get more views and faster?

This company, Rightster, grew to 250 people across twelve countries, and I was later recognised as the co-founder of what was by then the largest online video distribution company outside of America. We live-streamed Kate and William's wedding, over 2,000 fashion shows, the BRIT Awards and many other high-profile events. But the work still felt very manual – we had to work with thousands of websites in order to get the millions of views that our clients wanted. Even with an awesome seventy-strong tech team, I believed there had to be a way to escape endlessly copy-and-pasting the code that transferred the video from the cloud to the online website.

This is when I first heard about Data Science, Machine Learning and, eventually, AI. I thought it was a miracle. Sleepless nights could be a thing of the past – the machines could take over! But of course, it wasn't that simple. Charlie sent me on a mission to investigate, and I spent an entire year, and a significant amount of the company's money and my colleagues' patience, trying to harness this new technology. The dream was to use AI to get the right video to the right person at exactly the right time. This would then make them click on a link. We needed this to be done automatically, without having to predict which sites should embed the video, or which video to serve to which user. But, frustratingly, we didn't achieve the necessary financial results fast enough, and people lost faith. Back then, there were only a few companies who knew how to build AI systems, and it was nigh on impossible as a layperson to recognise the hidden gems. I learned the hard way that AI wasn't magic, and it would take a lot more than just a passion for it to work.

This experience shaped our next big leap. After Rightster went public and acquired similar businesses in the field it was time for us to leave. We agreed to dedicate our time to better understanding AI and learning what it meant for other companies. Lucky to be in London and surrounded by some of the best universities in the world we hosted dinners for big brains and debated the hottest topics of the day. It was clear that AI was a paradigm shift that was going to change the whole of society, not just our business. Now, this may sound dramatic but as you'll see later in the chapter on history this is not a new phenomenon or a novel revelation – AI has been around since the 1960s – but it was a concept to which the business world was only just waking up and it was the most exciting time. Our findings led us to setting up CognitionX.

CognitionX now employs over fifty people and provides a platform that connects businesses with questions about AI, or other emerging technologies, to the right expert in their own organisation, or thousands of experts in the wider community, who can provide them with answers and so help them navigate the complex AI landscape. For example, we've written reports for the Mayor of London's Office, where we analysed the AI ecosystem in order to help develop policy to encourage more companies to set up in the city. It's an honour to get to work with such incredible brains, be they from big corporates such as HSBC or from local government. Each year we hold a festival in London called CogX, where over 30,000 visitors from industry, government, civil society and academia come together to discuss how to forge a brighter future.

My passion for the AI community meant I spent every waking minute untangling the key questions the community faced. I cherished every opportunity to connect people with problems and in 2018 I was appointed by Matt Hancock and Greg Clark, two Secretaries of State in the UK Government, to assemble a team of top experts in AI to form a Council. The group is responsible for supporting the government of the day and its Office for Artificial Intelligence, currently helmed by Sana Khareghani, to ensure that future decisions about AI enable the UK to flourish and for all of society to benefit.

It's fair to say that I threw myself into the world of AI and tried to talk to every machine going. Whether in a meeting or with my friends, being interviewed on *BBC Breakfast*, I'd take every opportunity to explain why AI was the best thing to happen to humans – I was overly optimistic to say the least. But things began to shift when I realised that this technology would affect different people in different ways. I started to read widely and deeply on the topic, and I woke up to the reality that it could potentially have seriously negative effects for many women. This led me to gather women and host events looking at 'Why we needed Women in AI', in the mission to understand the scale and depth of the challenge I saw before us.

I was giving a talk a few years ago at an event hosted by the achingly cool *Riposte*, a magazine for women in creative industries. I looked around the room and realised I couldn't give my usual positive speech about AI and business. These women were at the top of their respective games, and I needed to explain why AI would matter to *them*. In fact, I welled up as it dawned on me how many women's careers could be destroyed by smart machines if they weren't aware of how AI was set to change all the rules. This book is one small step in hopefully arming those women – and you.

Getting Started

I've been fortunate enough to have a front-row seat as the world is transformed by new technology, and my aim here is to give you one too. I want to share with you the trials that are coming our way, as well as the potential that exists within AI for positive change. The way I see it, this new wave of technology could be a tsunami that knocks you down, or it could be the wave that we ride together to a brighter future. The moment I began to truly understand this, I knew I had to share what I'd learned about its possible risks as well as its rewards – and why it is that women could be more likely to suffer the negative effects. So it's important to stop seeing tech as 'boring', 'scary' or 'for someone else'. I'm not a scientist, engineer, developer or techie. It takes me a long time to understand technological ideas because they're mostly founded in complex mathematics. It was a really liberating moment when I realised that I didn't need to understand the precise inner workings of AI machines in order to understand the ramifications of this technology. All you need is to get a good grasp on how to adapt and thrive in this new world and what you can do to support others to do the same.

At this point, I think it's important to stress to you that the world of AI is rapidly changing all the time. It's impossible for me to condense the mass of compelling and complex ideas out there. Instead, this book is for all of us who need a good enough understanding to get by; not the intimate technical details. Consider it a patchwork quilt of sorts, an introduction to AI and an invitation to further learning. At the heart of this book is a chapter of interviews with a cross section of women who have informed my thinking. So grab a biro and get ready to add your own thoughts to these pages. Scribble across and circle areas you want to look up later. I often mention articles and academic papers – you can use a search engine to find out more about what they say. If you're reading on a Kindle or listening to an audiobook, maybe you can use voice notes – or even have an AI transcribe them!

Continue your learning with the further reading and watching at the back of this book, check the suggestions I've included, attend events and test new theories yourself. I hope *How to Talk to Robots* will give you the confidence to continue your journey of discovery and find your rightful place in the world of technology. This might well be the first book about 'robots' you have picked up, but I'm hoping it won't be the last.

KEY WORDS & PHRASES

to help you use this guide

Here are a few of the regular words and phrases you might stumble across while reading this book. Some are used widely, others are more niche. Don't feel as if you have to read this section before you get stuck into the chapters – it's just here for you to use as a reference any time you spot something that needs a quick definition. If anything else is unclear, why not ask Siri or other voice-activated assistants on your phone, who will use a combination of speech recognition and natural-language processing to help you out.

HOW TO TALK TO ROBOTS: I titled this book *How to Talk to Robots* because *How to Talk to an Artificial Intelligence System* wasn't quite as snappy, and I wasn't sure I'd get your attention. In most cases these AI systems won't have a physical 'body' so please don't confuse the two. Throughout the book, we'll keep referring back to this challenge of interacting with AI – all I really mean is how to engage with products, either physical or digital, that use artificial intelligence. And not all robots use AI either!

ARTIFICIAL INTELLIGENCE (AI): This is a tricky one because the definition of AI is hotly debated. At its most basic, AI describes a computer system able to perform tasks normally associated with human intelligence, like reasoning or learning. Within this wide definition of AI, there are two important distinctions.

1. **ARTIFICIAL NARROW INTELLIGENCE (AI):**
   Artificial Narrow Intelligence is an umbrella term for a system that can perform a single intelligent task. It is built for a single objective, such as automating image sorting, or text generalisation. This is the kind of artificial intelligence researchers are currently able to build.

2. **ARTIFICIAL GENERAL INTELLIGENCE (AGI):**
   Sometimes called superintelligence, AGI refers to the ideal realisation of the field of artificial intelligence, where machines would be fully able to learn, reason and act for themselves, all at once.

AFFECTIVE COMPUTING: An area of research in robotics and AI focused on thinking about and simulating emotions and feelings in machines. In addition to computer science, affective computing can draw on ideas in information studies, psychology, philosophy and others.

ALGORITHM: This is a messy one! An algorithm at its most simple is just any set of rules a computer follows to solve a problem. But when I talk about algorithms in this book, I am referring to AI algorithms, or the instructions the AI takes to perform its task. These are often also called models when the algorithm has already been trained with data.

BIG DATA: A way of describing extremely large datasets by using a computer to analyse and reveal the patterns, trends and associations in one set.

COMPILER: A compiler is a system that translates code into a mode that computers can process.

DATA: Is made up of individual units of information. Although the terms 'data', 'information' and 'knowledge' are sometimes used interchangeably, in the case of AI it's what people use to describe the information that an AI system is trained on or given to learn from.

DATASET: A collection of data pulled together for analysis or study.

DEEP LEARNING: Is a type of Machine Learning where the computing system mimics the structure of the human brain to learn from experience. This is why the structure of a deep learning system is often called a **NEURAL NETWORK.**

EXPERT SYSTEMS: A kind of computer system that uses databases of knowledge to make decisions, much like a human expert. This term arose in the 1970s, had its boom in the 1980s, but is used less often today.

FILTER BUBBLE: Personalised algorithms can create a 'bubble' effect so that a Web user only encounters information and opinions that reinforce their existing interests or beliefs.

GENDER DATA GAP: Caroline Criado-Perez uses this term in her book *Invisible Women* to describe the ways that data sets are often male-biased. This means the conclusions drawn from that data are less likely to reflect the needs and experiences of women, and women of colour in particular.

NATURAL LANGUAGE PROCESSING (NLP): A diverse field of study within artificial intelligence that asks questions about and helps machines process human language.

NEURAL NETWORKS: A layered structure of algorithms used in deep learning systems. This is why deep learning and neural networks are terms commonly intertwined when talking about how AI works.

MACHINE LEARNING: A kind of Artificial Narrow Intelligence where a machine learns to make decisions from data and experience, instead of it being explicitly told how to perform a task. Machine learning is called **DEEP (MACHINE) LEARNING**, when there are multiple steps in the decision-making process. There are three major types of machine learning: supervised, unsupervised and reinforcement.

STEM: This is a shorthand way of referring to the fields of science, technology, engineering and mathematics.

TRAINING DATA: This is the data that contains labels for different groups so the machine can learn the characteristics of each group. This almost always forms the basis for supervised machine learning.

1

WHAT IS AI?

And why should it matter to you?

Have you used a Snapchat filter to see what you'd look like as a baby? Is your inbox delightfully clear of spam emails? Are your photos on Facebook automatically tagged with the correct friends? Is the autocorrect on your phone so good you think it's psychic, or so absurd that you text it to your best friend anyway? Have you had a full-blown conversation in French using one of the speech-to-speech translation apps? Are you guided – or bitterly disappointed – by Netflix's film and TV viewing recommendations? If you answered yes to any of these questions, you are already interacting with artificial intelligence.

The AI that many people use every day is getting more accurate all the time. Take Google Maps. Originally, this app just got you from A to B. Then it helped you get from A to B with the least amount of traffic. Now it asks how busy your tube or bus ride was, then translates this information to let you know if there's usually only standing space at particular times. Then Google Places appeared, showing you inside buildings, restaurants and points of interest across the world. It can make recommendations for where you might like to visit. If you use a Google calendar to record your dentist appointment, it will automatically personalise your map to include where your dentist is. It'll even suggest when to leave your meeting in order to get to the next one on time. Last night, Google Maps dropped a pin when I parked my car to remind me where it was after I returned from dinner. Initially I thought this was spooky but then I was grateful because I always seem to forget. By the time you're reading this, who knows how much more advanced the simple map will have become. We'll discuss later the risks of relying too much on this new technology; we'll also talk about who has access to this technology and who does not (and why that matters). For now, if you do use a smartphone, just hold in your mind all the ways that your day-to-day is being affected by tech in the palm of your hand.

AI is automating everyday tasks at an unprecedented pace. As we acquire and learn how to use more new digital devices, we're also constantly upgrading our dependency on AI. Few people are oblivious to how this technology has firmly slotted into our lives, but most of us don't appreciate that the rules of this tech are constantly shifting and evolving. We're now living in an era where machines are taught to learn and adapt without human intervention, and this has some serious ramifications that we'll explore over the course of the next few chapters.

The many functions of AI will continue to impact every aspect of our daily routines, but as well as helping us to avoid traffic or introducing us to new music, AI systems will detect disease, reduce energy consumption, decide which of us receives an approved loan, power vehicles to be autonomous and self-driving and both inform and control our news and advertising feeds. AI has the potential to unlock a future where humans live longer, healthier, happier lives. It could change the nature of much work: taking over many repetitive, boring tasks and freeing humans up to spend time on creative, fulfilling projects. This could dramatically change common conceptions of how much of our lives we need to spend on work, allowing us all to spend more time with each other and on our relationships, or on whatever it is that makes life worth living for you.

But there is a much scarier alternative. The flipside to this technology is that it could make life a lot worse for a lot of us, and especially the most vulnerable: it could widen the poverty gap, further increase inequality, reduce diversity and re-entrench many of the structures that keep some people down, no matter what they do. Anyone who tells you otherwise is not telling you the whole story.

The reality of AI, in its current state, is that it adopts the truths of its creators: humans. AI-driven machines learn from data that humans feed into their systems, meaning that they can also learn the social norms that many of us are desperately trying to escape or eradicate. For example, what if they are programmed to accept the existing pay gaps, or the idea that a woman's place should always be in the home? If misogyny and unconscious or conscious bias is codified into the next wave of technology, we are all exposed to a less fair, less equal future. It gives me the shivers just thinking about it. So, although AI has enormous potential to improve our lives, there is the risk that rather than empowering women it may continue to compound existing stereotypes. And, as we're going to see below, the evidence that we have suggests that the risk is already becoming the reality. We have to protect our rights and fight against oppressive gender constructs becoming codified, because if they are, then whatever progress women have made over the past century could very easily be wiped out. We need to do so much more than become merely competent responders to AI.

The first step here is to accept that to thrive we must learn to live and work alongside machines. I don't want women to be more at risk of losing their jobs to a machine because there wasn't enough material out there to prepare them for using it. But don't worry, as you'll read time and again from some of the women interviewed here, this does not mean we have to become coders, statisticians, designers, or engineers. On the flipside, we certainly *can* do all these things and in fact, many of the pioneering early coders were women.

No one will be unaffected by AI. Are you at school deciding what work experience to do? Are you in university and making plans to enter the world of work? Are you in an office or in retail? Maybe you work in a hospital as a nurse or a porter or a doctor. Or are you a banker, an accountant, an advertising executive? Are you a part of the gig economy? AI technology, as you perhaps know all too well, is already permeating your workplace. The challenge now is to work out how to make sure it helps you rather than undermines you.

It's widely publicised that only 13 per cent of engineers and data scientists working in the West are women, and so I hope some of you go on to change this shocking statistic. But what this book is really about is inspiring you to get interested in tech, feel comfortable working alongside it, leverage it, use it, be enabled by it, know how to be heard and have a stake in how technology is built and deployed. It's a pragmatic guide for the uninitiated.

There are ways that you can be instrumental to the building of the AI systems. Just think about how many different steps there are when building a product. Companies are going to need people who love languages to give AI a voice; historians and philosophers to give AI context; designers and artists to give AI personality and an interface; and product managers to ensure the AI is fit for purpose. The number of non-technical roles will become crucial as we try to build machines that think and act like humans. AI should be democratised not professionalised. All this power should not lie in the hands of the few, nor should it stay only in the hands of men who currently make up the dominant percentage of the tech workforce.

OK, sounds good, but what exactly is AI?

In order to answer this question, I called on **Karen Hao**. Karen is a journalist, storyteller, engineer and for the past few years she's been the woman who has explained complicated concepts to me via the *MIT Technology Review* magazine. She has made what hundreds of other people have tried to explain to me before just click into place. She's always been good at finding novel ways of communicating: as well as an AI expert, she's also passionate about the environment and as an undergraduate, launched a fashion show entirely made of rubbish! She's a woman after my own heart in more than one way.

What are the fundamentals of AI that everyone should understand – and why does it often seem confusing?
Well, in the broadest sense, AI refers to a branch of knowledge that strives to recreate human intelligence within machines. In

Karen Hao

the ideal realisation of this goal, such machines would be able to learn, reason and act for themselves, mimicking the ways we as humans, or a pet dog, might do so. They would take in information from the rapidly evolving world, process it and then figure out how to respond based on prior experience – and they would supposedly be able to do so much faster and on a far greater scale than any individual human could.

This is the dream of artificial intelligence: to make these super capable machines that can put their 'minds' together with ours to solve some of the world's most complex problems: climate change, poverty, hunger – things we haven't been able to wrap our heads around on our own. I like to think of this dream as Janet from NBC's comedy series *The Good Place*. She's a kick-ass, fully autonomous and highly intelligent agent that helps her human counterparts be better versions of themselves.

Part of the reason why AI gets confusing is because the term can often feel like it refers to two completely different things. We now know one of them – the vision of the field, which evokes something closer to what we see in science fiction. But as you've already intuited, today's AI systems are nothing near the clever, autonomous agents that I just described. They're simpler and less capable, able only to perform specific tasks such as adding dog ears onto your Snapchat photo, ranking your content on Facebook or recommending new songs on Spotify to match the genre you like.

In the field, these two versions of AI have been given different names. The dream is often called 'artificial general intelligence', or AGI, while the reality is sometimes referred to as 'artificial narrow intelligence'. People usually call artificial narrow intelligence AI. But these definitions are often combined into one in popular culture, causing people to think that the AI we have today is far more advanced than it really is.

There is another element that complicates the whole thing further. Because AI researchers are constantly pushing the boundaries of the technology, the definition of present-day AI also changes over time. What might have been considered AI thirty years ago is no longer really considered AI today. And what we know as AI today has only been the working definition for less than a decade. Certainly, among experts there's a considerable amount of debate about what constitutes AI and what it will be capable of in the future. In fact, one of the fiercest debates is about whether AGI is even possible!

When you add all that up, it goes without saying that the notion of AI is constantly being tweaked, debated, probed and refined. Don't let that overwhelm you. Instead, take it as an invitation: what AI is and where it's going is ultimately shaped by people. That means you can have a role in influencing the technology, which I hope you find as hugely exciting as I do.

So what do you think the most common misconceptions are about AI?

One of the biggest things that confused me when I first began covering the field is the difference between robots and AI. The two are clearly interrelated, yet they are not the same thing. As it stands now, AI specifically deals with software. Robotics, by contrast, deals with hardware. Think of it as your brain versus your body. Your body relies on instructions from the brain to move, but your brain also relies on your body to experience and sense its surroundings so it can learn about the world.

Of course, as mentioned in the introduction, AI doesn't always come physically embodied in a robot. More often than not, it exists as a hidden algorithm on your favourite websites or apps. Similarly, robots aren't always powered by AI. Instead, their actions could be dictated by software that executes a series of hard-coded rules. In those instances, the robots can neither learn nor adapt to unexpected circumstances, so they are usually confined to perform rote tasks in unchanging environments such as a manufacturing floor.

When AI and robots combine, that's when the interesting stuff happens. It's no longer only about efficiency but breaking new ground as well. Self-driving cars, for example, are a product of this union.

What happens if we rely too much on AI?

The most popular products we use are developed and implemented by tech corporations with profit motives. That's why social media and streaming platforms can be so addictive: the machine-learning algorithms are all pushing us to stay on these platforms for just a little longer. That's also why ads can sometimes turn predatory: the algorithms get so good at knowing our weaknesses that we end up spending more money than we should have. As algorithms learn more and more about you, implications regarding privacy become a serious concern.

It's important, therefore, every once in a while to think through how algorithms are impacting your life. What do you like about how they've changed it? What do you not? Ultimately, you are in control of the way you interact with them. You could choose to abstain from them entirely: Facebook, for example, gives users the option to turn off automatic photo-tagging. Or you could learn to hack it so that it does different things: if YouTube is showing you too many similar videos, purposely watch some radically different ones to reteach the algorithm about what you want.

What's your one piece of advice for the reader?

Command algorithms; don't let them command you! And you're already one step ahead of the game because you're reading this book.

As Karen has explained, AI has been around for a lot longer than it first appears. The next chapter takes a closer look at the history of AI. It might not be what you're expecting . . .

2

A POTTED HISTORY OF AI

One of the many fascinating aspects of AI is that although it still seems super futuristic, its roots lie in classical thinking. Since we humans have always been partial to making things easier for ourselves, we've long been pretty adept at designing machines to increase efficiency. Combine that disposition with some killer mathematics and it's little wonder that AI has been a twinkle in our eye for quite some time.

The quest for artificial intelligence as we know it began over seventy years ago with the idea that computers would one day be able to *think as humans*. As so much of the history of AI stems from early computers, we'll start off by looking at where it all began. You'll soon notice that early AI was influenced by many different disciplines – yes, mathematics and engineering play their part, but so do biology, game theory, psychology and philosophy. AI is not just the domain of computer scientists. It has always been, and will continue to be, an arena that explores what it means to be human and how we live our lives.

	205BCE	Antikythera mechanism
1930s	**1930**	Alan Turing discusses the idea of computers that can think
	1939	WW2 breaks out and funding pours in for computation projects to help the war effort
1940s		
1950s	**1950**	Turing publishes paper outlining the Turing Test. Science Fiction boom: *I, Robot* is published
	1956	John McCarthy, Marvin Minsky & others first confer on intelligent computers 'Artificial Intelligence' coined
1960s	**1961**	Marvin Minsky publishes 'Steps Toward Artificial Intelligence'. Unimate first industrial robot on General Motors assembly line
	1969	The ARPANET (now known as the Internet) first goes live
1970s	**1970**	Japanese robot Wabot is developed
	1973–80 the first AI Winter	
1980s	**1980**	Development of expert systems, ending the AI winter
	1982	Research on Deep Learning begins from Hinton, Rumelhart & Williams
	1987	DARPA funding ends
	1988–93 the second AI Winter	
	1989	Deep Thought beats Garry Kasparov at Chess
1990s	**1991**	Tim Berners-Lee and team invent the World Wide Web
	1997	Deep Blue beats Garry Kasparov at chess
2000s	**2002**	Roomba released
	2004	DARPA Grand Challenge for autonomous vehicles
	2005	Recommendation technology tracking web activity brings AI to marketing
	2008	Google introduces first speech recognition software into devices
2010s	**2010**	IBM Watson beats human at Jeopardy
	2011	Siri released
	2012	Hinton deep learning breakthrough
	2014	Alexa released, the first machine passes the Turing Test
	2016	DeepMind AlphaGo beats master Go player Lee Sedol
	2018	Alibaba system AI claims; OpenAI GPT-2 release deemed too dangerous
	2019	DeepMind's AlphaStar defeats almost all human players at StarCraft II

1836 Ada Lovelace works with Charles Babbage on the Analytical Engine

1930 Klara and **John Von Neumann** begin work with Turing

1930s

1940 Joan Clarke and Turing, plus a team of coders crack the Enigma code at Bletchley Park

1943 Klara Von Neumann helps train **Jean Bartik**, **Betty Holberton**, **Kathleen Antonelli**, **Marlyn Meltzer**, **Ruth Teitelbaum**, and **Frances Spence** on the ENIAC

1940s

1952 Grace Hopper develops the first compiler while working on Mark I computer with John Von Neumann

1958 Phyllis Fox and John McCarthy develop LISP

1950s

1961 Dorothy Vaughan begins to teach herself and her NASA team at FORTRAN

1965 DENDRAL project, **Georgia L. Sutherland** is key writer for the program it ran on (in LISP)

1967 Hopper wins the Computer Science Man of the Year Award

1960s

1972 Vaughan retires from NASA, where she was leading the Electrical Computing Unit

1972 Karen Spärk Jones inverse document frequency paper revolutionises NLP

1975 Elizabeth Jake Feinler domain name definition for ARPANET

1970s

1980s Radia Perlman develops Spanning Tree Protocol setting the standard for internet traffic

1989 Wendy Hall begins work on Microcosm

1980s

1990s Leslie Kaelbling begins working on the concept of reinforcement learning

1997 Kaelbling wins IJCAI award for her work on reinforcement learning

1990s

2000 Cynthia Breazeal develops KISMET, the robot with feelings

2007 Fei-Fei Li develops ImageNet, ushering in a new era in computer vision and image analysis

2000s

2017 Cathy O'Neil publishes *Weapons of Math Destruction*

2017 Li creates AI4All

2017 Dame Wendy Hall & Jerome Pesenti release UK government *AI review*

2010s

2017 AI Now Institute is founded by **Kate Crawford** and **Meredith Whittaker**

2018 Joy Buolamwini and **Timnit Gebru** bring attention to racial bias in AI

AI's interdisciplinary nature means that there are many ways to trace its history. This chapter is one way and is not exhaustive. Instead, I wanted to focus on some of the major moments and shine a light on some of the women who've been so chronically overlooked along the way. There is a long tradition of women translating or communicating science, partly because of being excluded in some way from the mainstream of conducting it. In fact, I've searched far and wide for all the women who played a role in the development of AI, but many were simply never recorded in the archives. This is why I've called this chapter a potted history: because I've chosen the accounts that most inspired me. Consider this just one way of telling a very complicated story. A historian of science at the University of Cambridge, Patricia Fara, summarised this perfectly when she said to me: 'Broadening what counts as science's history entails recognising and crediting women's involvement.'

I've ordered this chapter as a timeline of individuals, but it's really important to note that the history of AI, as any history of innovation, is never that straightforward. The history of computing in particular is often documented as a series of male geniuses appearing one after another, which simply isn't the case. So often when history is told as 'a series of geniuses', women, particularly women of colour, are erased from the narrative. Among the many reasons for this is the big one: traditionally, the people writing history are the same people who hold the power – white men. When new inventions or ideas shake up our way of thinking or doing, it's always a result of many people working together, forming communities and pushing things forward – not just one individual, inspired though they may be. So please remember as you read this that it's networks of people, not individuals on their own, who've brought us to where we are today.

Pre-Twentieth Century

The Antikythera Mechanism (205 BCE)

Mechanical machines have been performing complex functions for much longer than you would think. Salvaged from the depths of the Aegean Sea, the Antikythera mechanism has been described as the world's oldest computer. This strange thing has thirty-odd gearwheels and countless astronomical inscriptions, and it dates to 205 BCE – over 1,200 years before mechanical clocks first appeared in Europe. International experts have decided that it must have been part of an intricately engineered machine for crunching the mind-boggling mathematics needed to model the positions of the sun and moon. The fact that it not only monitored but also seemed to have forecast solar cycles is of course exciting for astronomy enthusiasts, but it also makes it part of the long history of artificial intelligence. Experts still don't know its exact purpose – it could have been a teaching tool, or used for navigation, but I like to think that it was an early version of the Clue app, tracking women's menstrual cycles. The likelihood of this is, sadly, zilch. As you'll see, technology developed in patriarchal societies rarely takes the needs of women into account.

The Analytical Engine & Ada Lovelace (1815–1852)

The partnership between Ada Byron, Countess of Lovelace, and Charles Babbage in the 1830s was perhaps the true birth of computer science.

As a child, Ada Lovelace was taught mathematics by her mother Lady Byron, who hoped to keep her well away from the poetry (and temperament) of her father Lord Byron. Babbage was the inventor of a machine he called 'the Analytical Engine', an early computer designed to complete mathematical functions. When Lovelace first met Babbage, he asked her to translate into English an account of his Analytical Engine written by the future prime minister of Italy. She made copious notes on the paper and proposed an algorithm for the Analytical Engine to calculate a sequence of Bernoulli numbers that blew Babbage away. This algorithm is arguably the first true piece of computer code. Though it was a century before the dawn of the computer, Lovelace went on to imagine a machine that could be programmed to follow instructions. But as well as calculating, it would also create. She anticipated a machine that 'weaves algebraic patterns just as the Jacquard loom weaves flowers and leaves.'

Although the computer she wrote about was never built, it was Lovelace's imagination, application and appreciation of maths and mechanics that earned her the title of first computer programmer and the epithet 'the Enchantress of Numbers'. It's exactly this kind of imagination that enables researchers, academics and entrepreneurs worldwide to push boundaries and make new discoveries with artificially intelligent machines today. I love knowing that it was Ada Lovelace, a woman well before her time, who sparked this line of enquiry.

Early Twentieth Century

Alan Turing (1912–1954) & Joan Clarke (1917–1996)

Alan Turing was an English mathematician and computer scientist and is now considered the father of theoretical computer science and artificial intelligence. Turing studied mathematics at Cambridge, and it was there that he first started thinking about the possibility of an intelligent computer. In the 1930s, Turing went to the USA and worked with John and Klara von Neumann – remember their names, we'll come back to them shortly! – who were also fascinated by the idea of possible computer intelligence. Shortly before the Second World War broke out, Turing returned to the UK with everything he'd learned and took up a position with the Government Code and Cypher School. You might recognise his name from the 2014 film *The Imitation Game*, a dramatisation of how during the Second World War he cracked the Enigma Code, which was used by the Germans to send commercial, diplomatic and military communications, and so helped the Allies to defeat the Nazis. But he didn't work alone. At their base, Bletchley Park, Joan Clarke rose from clerical work to deputy head of operations in Hut 8, becoming its longest-serving team member. To be promoted and receive her initial pay rise, Clarke's title had to be changed to 'linguist' as the Civil Service had no protocols in place for a senior female cryptanalyst. Clarke was tasked with breaking navy ciphers in real time, resulting in almost immediate military action. The secrecy of Bletchley Park means that many of her achievements remain unknown today.

What we do know is that the Bletchley Park code-breaking operation was made up of nearly 10,000 people and about

75 per cent of these were women. Very few women there have been formally recognised as cryptanalysts working at the same level as their male peers. Thankfully, in more recent times, Mavis Lever, Margaret Rock and Ruth Briggs have also been named. I like to think I would have been friends with Sarah Baring, as she combined her work for *Vogue* with that of her duties as a linguist in Hut 4 at Bletchley Park.

As Professor Sue Black, the woman who campaigned to keep Bletchley Park open, explains, 'the lifeblood of some of Britain's bravest and most inspiring citizens pulsed through Bletchley's veins at the most crucial turning point of the war. Here, thousands of men and women contributed to the effort that saved our nation and inspired future generations with their work in the fields of computing and technology.' What's great is that you can now go and visit these huts and put yourself in those women's shoes.

ENIAC, John & Klara von Neumann (1903–1957 / 1911–1963) & Jean Bartik (1924–2011)

Around the same time in America, husband and wife team John and Klara von Neumann were using their respective skills to build early computers to help the war effort. Despite only having high school maths, Klara secured a wartime job coding with Princeton's Office of Population Research. While her husband went to work on the Manhattan Project (war-time research on nuclear weapons), Klara become head of the Statistical Computing Group at Princeton until the end of the war. In peace time, she continued to program for many of the earliest computers, including the ENIAC, and trained others to do the same.

Meanwhile, ENIAC, the first digital computer, was being built in America for the war effort by John Mauchly and J. Presper Eckert. This innovative machine was, again, powered and programmed by a team of women. The original ENIAC wasn't a computer like we have today, ones that stores program instructions in electronic memory. Instead, think of it as a machine with a series of huge plug boards rigged up to thousands of wires that needed to be manually pulled in and out. Needless to say, it was laborious.

ENIAC was set up to calculate ballistic trajectories for weapons, but it was the hard work and pluck of the original programmers, Jean Bartik, Betty Holberton, Kathleen Antonelli, Marlyn Meltzer, Ruth Teitelbaum and Frances Spence, who got this beast going. What's more, all these women taught *themselves* how to operate the 150-foot tall machine as the engineers had no time for programming manuals or classes. They wrote the program by looking at ENIAC's logical and electrical block diagrams, and created their own flowcharts and programming sheets, navigating the hundreds of wires and thousands of switches to place it on the ENIAC. These women were drawing on their mathematical ability, but I think the extra skill was their determination to get these machines to work for them.

Jean Bartik did once say that the ENIAC was a 'son of a bitch' to program, but these women were devoted to their work. On Valentine's Day in 1946, Bartik and Betty Holberton were up late working on the machine ahead of a big press conference the next day. Reflecting on that evening, Bartik wrote 'Most people consider Valentine's Day a romantic day, but we never gave a thought that evening of romantic dinners. What we were thinking of was an all-important demonstration we were to run the next day for the world.' The day came but no one was told about the work of these women. The lab only introduced the ENIAC's hardware inventors – all men – to the press. Photos from the event show the women's faces snapped in the background, uncredited. After the big reveal, there was a celebratory dinner with many invited guests. The women were sent home. It wasn't until 1985 when Kathy Kleiman, an undergraduate programmer at Harvard, was searching for female role models that the story of the women of ENIAC was rediscovered. They had to wait until they were seventy years old to be recognised as the world's first computer programmers. And yet we still benefit from technologies that they helped to develop: Klara von Neumann, for example, applied her understanding of ENIAC technology to make huge breakthroughs in weather forecasting.

The Turing Test

These wartime computers proved that machines could be more powerful than first imagined. In 1950, at the dawn of the digital age, *Mind* published Alan Turing's seminal paper 'Computing Machinery and Intelligence', in which he posed the question, 'Can machines think?' Instead of trying to define the terms 'machine' and 'think', Turing outlines a different method derived from a Victorian parlour amusement called the imitation game. This later became known as the Turing test, and it demonstrated how a human may be unable to distinguish machine from another human being.

The Turing test involves three participants: a computer who answers questions, a human who answers questions and another human who acts as the interrogator. Using only a keyboard and a display screen, the interrogator will ask both human and computer a series of wide-ranging questions to determine which is the computer. If the interrogator is unable to distinguish the human from the computer then it is considered an intelligent-thinking entity and has passed the test.

I subscribe to the belief that the Turing test doesn't actually assess whether a machine is intelligent, more that we're willing to *accept* it as intelligent. As Turing himself said in 'Intelligent machinery', a report he wrote for the National Physical Laboratory, 'The idea of intelligence is itself emotional rather than mathematical. The extent to which we regard something as behaving in an intelligent manner is determined as much by our own state of mind and training as by the properties of the object under consideration.' Turing's paper had a huge impact in a variety of fields as questions about machine intelligence snowballed into those about a machine having emotions or soul. It prompted further lines of enquiry of philosophy, engineering, sociology, psychology and religion.

In Chapter Seven, we'll explore how, just as Turing predicted, modern-day AI products are built to kid you into thinking they are real humans. You'll find some pointers about how to develop the skills needed to better spot the difference!

Grace Hopper (1906–1992)

Meanwhile, back in America, John von Neumann was developing ever more complex computers, like the MARK 1. One of this project's first programmers was Grace Hopper. As well as being in the navy reserves during the Second World War, Hopper was excellent with numbers, and she used her PhD in mathematics to become a ground-breaking programmer.

It was Grace who came up with the idea of including English words in computer programming language, rather than the complicated numbers and symbols that had, until then, been used as commands. At first, her mostly male colleagues scoffed and thought it was impossible. But Grace persisted, determined to make programming more accessible. She's been quoted as saying, 'If it's a good idea, go ahead and do it. It's much easier to apologise than it is to get permission', and I can't help but admire this attitude from a woman who knew that some people would see her gender as a reason to hold her back. In 1952, she successfully released her first **compiler** (a way to translate between human language and computer language) called A-O. This idea of the compiler is what many coding languages still use now – a mix of numbers and letters compiled into a version that the computer can make sense of. Then, in 1959, she led the development of one of the earliest coding languages, COBOL.

What I think is so critical about Grace's role in AI is that she was trying to make programming *more* accessible by effectively being a translator of computer language, removing it from the clutches of the few men who knew the program language. She considered those programmers as 'high priests' who regarded themselves as the gatekeepers between ordinary people and computers. This meant that while she was making all these incredible breakthroughs in programming, she was not popular among her male peers. Later she said that 'Someone learns a skill and works hard to learn that skill, and then if you come along and say, "You don't need that, here's something else that's better", they are going to be quite indignant.' But once more, she persisted. Hopper also went on to become one of the key people building and developing the standards for other programming languages like FORTRAN – paving the way for my own personal hero Dorothy Vaughn (more on her in a moment).

Hopper played a critical role in computation and AI. In 1969, she received the Computer Science Man of the Year Award from the Data Processing Management Association, to which was added throughout her life a whole host of other medals and awards that no woman had ever received before. She was even featured in a 1967 issue of *Cosmopolitan* magazine about women in computing – no mean feat for that time.

LISP, John McCarthy (1927–2011) & Phyllis Fox (1923–)

Although Turing was already asking 'can computers think?', it wasn't until 1955, two years after Turing's untimely death, that a young computer scientist, John McCarthy, coined the term Artificial Intelligence.

A year later, McCarthy brought together a group for the first conference on AI, and this is really when the field as we understand it began to take shape. Sadly, this is also the point at which the AI world became overwhelmingly male. The conference was organised and attended only by men. (It's not that women were not still making waves in this new AI arena – they were – but they were often sidelined and rarely recognised.)

In 1958, McCarthy created the LISP computer language, which is still used. But what many people don't know is that the first interpreter and manual for LISP was written by a woman named Phyllis Fox. As Fox humbly commented in an interview with the

Society for Industrial and Applied Mathematics, 'Now, this was not because I was a great LISP programmer, but they never documented or wrote down anything, especially McCarthy. Nobody in that group ever wrote anything down. McCarthy was furious that they didn't document the code, but he wouldn't do it, either. So I learned enough LISP that I could write it and ask them questions and write some more. One of the people in the group was a student named Jim Slagel, who was blind. He learned LISP sort of from me, because I would read him what I had written and he would tell me about LISP and I would write some more. His mind was incredible.'

I've often been the only person in the room who would just get on with it and write something down to preserve it when the men wouldn't. Phyllis Fox is an inspiration, not only as a woman coder, but also for reminding us of the importance of translators, preservers and teachers to the history of AI.

The 1960s and 1970s

DENDRAL & Georgia L. Sutherland

In the mid-1960s, a problem-solving AI project called DENDRAL was set up. It tried to use ideas from artificial intelligence to help chemistry labs and researchers identify unknown molecules by storing information about what they did know and then analysing those they didn't in order to look for patterns and similarities. DENDRAL ran for decades, and a lot of the software and theory it used ended up being the basis for expert systems in the 1980s (more on that later.)

If you research DENDRAL, the names of four men (Edward Feigenbaum, Bruce G. Buchanan, Joshua Lederberg and Carl Djerassi) are repeatedly affiliated with the project. But it was a woman, Georgia L. Sutherland, who, as you can see from the published notes, wrote the program in LISP and was in fact the lead author on many of the papers.

Dorothy Vaughan (1910–2008)

Remember Grace Hopper and her work with FORTRAN, the programming language she helped design and build? Well, not only did FORTRAN become an incredibly important language for businesses and industries, it also played a huge part in the Space Race at NASA.

If you've watched the film or read the book *Hidden Figures*, which I suggest you do if you haven't, you might already be familiar with the name Dorothy Vaughan. Vaughan worked at NASA during the segregated Jim Crow years, where women were employed as 'computers', calculating complicated maths formulae by hand. When Vaughan saw that NASA were about to implement new technology that would do the jobs of her team faster, thereby making them obsolete, she decided to teach herself FORTRAN. She got the system working and then passed this knowledge on to her team, teaching them all that she'd learned. In doing so, she was able to make sure these women were skilled enough to be re-deployed to new jobs working on the most important part of the system.

Vaughan was the first ever African-American woman to be promoted to supervisor for the west computer group, the segregated female computing team. She led the electric computing department there and stayed at NASA until 1971.

Rather than feeling threatened by the machines, Dorothy embraced them, and with the confidence and foresight to predict the next wave of technology, she ensured that she and her team were primed and had the skills to keep their jobs and stay ahead of the machines.

Unimate

Tech wasn't only being used for space. In 1961, Unimate was the first robot built for manufacturing and was used to automate repetitive tasks at a General Motors plant. This was the first time the idea of stored memory (early AI) was used in industry, instead of in a lab or a research setting. It prompted a wave of robots being built through the 1960s and 1970s, notably the WABOT in Japan, which was the first robot built to resemble a human.

Kathleen Booth (1922–) & Karen Spärck Jones (1935–2007)

The late 1960s and early 1970s was a real boomtime for AI, and Karen Spärck Jones was part of it. Instead of focusing on the growing field of coding, she taught computers to understand the human language. This is called natural language processing (NLP) and it is critical for the idea of computer intelligence. Before Jones, little work had been undertaken to get computers to understand spoken or written English, although as early as the 1950s a brilliant woman named Kathleen Booth first imagined some of these concepts. Jones, however, was responsible for one of the biggest breakthroughs in NLP with something called IDF (inverse document frequency), a way of calculating how important a term is in a text. Her 1972 paper laid the groundwork of IDF, and this became the basis for almost all the search engines that we use now.

Like so many of the women featured in this chapter, what I admire and think is so cool about Jones is that she was a self-taught programmer – her background was in philosophy and history. I can't help but think that her immersion in these disciplines sparked her approach to computer science, which focused not only on how to make computers understand humans, but also its social impact.

She also came up with the phrase: 'Computing is too important to be left to men.' Karen Spärck Jones, I salute you!

The First AI Winter

Okay, so I've rounded up some of the prime events that led towards the development of AI, but in truth it's not as streamlined as it might appear on the page. Once the quest for artificial intelligence was truly underway, expectations snowballed, making the lack of advancement and results by the end of the 1970s dispiriting. Problems with progressing centred around two basic limitations: there was not enough RAM memory, making the processing speeds poor. This led to a lack of data suitable to train the machines. Progress slowed and research money was pulled. What followed came to be known as the 'AI Winter', which we can roughly date as beginning in the mid-1970s and ending in 1980.

The 1980s

Expert Systems

The emergence of expert systems in the 1980s went some way to address this. An expert system is a computer system that has an in-built decision-making ability. They work by using pre-programmed bodies of knowledge to reason their way through complex problems. These were the first truly successful forms of AI software and focused on narrow tasks, such as playing chess.

All around the world, expert systems were developed and quickly adopted by competitive corporations – just like Georgia Sutherland and DENDRAL. This meant that the primary focus of AI research turned to accumulating comprehensive quantities of knowledge from various experts. It didn't take long before AI's commercial value started to be realised, and investment into it began to flow again.

Deep Learning

Alongside the expert systems in the 1980s, another method of AI was being developed. Deep learning is a type of machine learning that tries to mimic the way humans learn through experience. You'll learn more about all the technical parts of this in the next chapter. Some of the core ideas from machine learning – the idea that a computer can learn to make decisions from identifying data patterns – can be traced to those early days in the 1950s, but deep learning brought machine learning to a new level. It was first developed by mathematical psychologist David Rumelhart and his doctoral advisor, physicist John

Hopfield. They were later joined by Geoffrey Hinton, and it was their idea to train machines by mimicking the design of the brain with neural networks. Deep learning is still best practice for much of AI in the twenty-first century.

Deep Thought

Around this time, many people were still engrossed in how AI could work *with* humans, rather than just for them. One of the most common ways to explore this was through games. Engineers and computer scientists wanted to know: could a computer understand context sufficiently enough to play a game with a human? One of the most famous examples of this is Deep Thought, a chess program.

In 1988, Deep Thought beat its first grandmaster (someone ranked in the highest level of chess playing), marking a real turning point for AI. Within a decade, IBM's chess playing application, Deep Blue, beat world-reigning chess champion, Garry Kasparov. The machine was capable of evaluating up to 200 million positions a second. But could it think strategically? The answer was such a resounding YES that Kasparov believed a human being had to be behind the controls. He also thought that the human was, of course, a man. In fact, Kasparov famously said, 'Women, by their nature, are not exceptional chess players: they are not great fighters.' Little did he know just how many women had already contributed to the development of the machine that had just shown him up.

By 1985, a billion dollars had been spent on AI. New, faster computers convinced American and British governments to start funding AI research again. But with the widespread introduction of desktop computers, it soon became apparent that expert systems were too costly to maintain. In comparison to desktops, expert systems were difficult to update and still didn't learn. In 1987, the Defence Advanced Research Projects Agency (DARPA) concluded AI would not evolve sufficiently and instead chose to invest its funds in projects they decided would yield better results. A second, though this time much shorter, AI Winter began.

Leslie P. Kaelbling (1961–) & Cynthia Breazeal (1967–)

In the late 1980s and 1990s, while AI had been toppled from its position as the golden child of the tech world, two women were nonetheless quietly making strides in developing it.

Leslie Pack Kaelbling is widely regarded as one of the leaders of AI in the 1980s and 1990s. She worked on the idea of **reinforcement learning**, which is one of three key areas of machine learning (the other two are supervised and unsupervised machine learning – more on this in Chapter Three). This was a departure from deep learning because machine learning is concerned with a computer learning from a set of data, whereas reinforcement learning learns from trial and error. It was Kaelbling who figured out how to use this for improving robot navigation – going on to win awards and edit leading journals in AI.

Meanwhile, Cynthia Breazeal was following a different line of enquiry. She wanted to know if machines could *feel*. This was called **affective computing** and Breazeal put this to the test by building a robot named KISMET. It was designed to recognise and simulate emotions and was one of the first ever robots able to demonstrate social and emotional interactions with humans. KISMET can perceive a variety of our emotions, and even gets uncomfortable when people get too close to it. Breazeal founded the JIBO company and continues to work on personalised robots today. Although Breazeal comes from a computer and engineering background, she also works in media, education and psychology.

Meanwhile, Going Online

You might be wondering at what point the Internet first appeared in the narrative of AI. The history of the Internet is a book in and of itself (see my notes on Claire L. Evans' *Broad Band* in Chapter Eight), but the connections between the Internet and AI are difficult to break down since they're now so intertwined. Think about how often people building algorithms have to contact other people to problem-solve, innovate and collaborate. But one way to look at this link is to consider how the Internet gave way to the collection of big data.

At its most simple, big data is the term given to datasets that are so big and complex that trends or patterns cannot be found without computational methods. A good example of this is how Facebook networks can be not only your friends, but also your friends of friends. Big data is critical for AI because it needs huge amounts of data to learn, and these datasets are often made available through the Web. It's not that the Internet is big data per se, but rather it's the way that big data has become so ubiquitous. When you're browsing online, accepting cookies usually means you're allowing a webpage to collect your data. So, although you are just one data point, the points from everyone who visits that page are pooled together into a dataset so large it can only be handled by computers.

This can all be traced back to a moment in 1969 when two devices first connected to each other over something called ARPANET. This is the earliest ancestor of what we now know as the Internet.

Before we go any further, I think it's important to make a quick distinction between what is meant by the Web and the Internet. It's pretty common to use them interchangeably, but technically the Internet is a collection of global networks while the Web is one way to access the information stored there. The Web can be traced back to 1991, but there are earlier examples of the same kind of systems from the eighties.

Jake Feinler (1931–) & Radia Perlman (1951–)

Computer scientists have been interested in the idea of computers communicating with one another since the 1950s. This led to a development of networks – LAN and WAN, the names of which you might recognise from using your desktops or laptops when connecting to WIFI. I find it helpful to think of these networks like the underground roots of trees that spread out and connect to each other, sharing water or sugar and other chemical, hormonal and electrical signals. Computer networks work in the same way, but they share digital information instead. LAN is a local area network where computers are linked to each other in a single building or location. WAN is a wide area network where multiple LANS connect to each other across a larger geographic space. In the 1980s, both these systems were riddled with issues, but two amazing women set out to conquer them.

Elizabeth Feinler, known as Jake, solved a huge problem with
WANs. One of the most famous and successful early WANs
was ARPANET, which started out linking several university and
military LANs to one another. But in these early days, WANs had
great difficulty navigating – imagine having no search engines,
no way of knowing how to find the person or organisation you
need. Along with her team, Jake solved this and progressed
the advancement of the Internet. She went on to be part of the
naming authority for the Internet, devising and developing top-
level domain names, including .com, .us and .gov.

Next up is Radia Perlman, who devised something called the
Spanning Tree Protocol. It allowed computers on LANs to connect
to more complicated networks without looping back along a
path, which would have ultimately broken the connection. This
protocol still forms the basis of many networks today.

Margaret Boden (1936–)

Margaret Boden is a Research Professor of Cognitive Science at
the University of Sussex and an expert in the intersection of AI,
psychology and philosophy.

She published her first article about AI in 1972, and has since led
the way when it comes to researching about the intersection of
computer science and the human mind. Stressing the importance
of celebrating creativity, she continues to be a powerful voice
advocating for the necessity of interdisciplinary research in
AI, which means she is essentially a trailblazer for how I think
about interacting with, and getting involved in, AI. She also thinks
through the possibilities of machine creativity, with a special
focus on what AI might mean for art in the twenty-first century.

Wendy Hall (1952–)

Dame Wendy Hall is one of the people who worked on a pre-
Web hypermedia system. A hypermedia system links together
different types of information like text or images in a non-linear
way. Wendy led a team of computer scientists who collaborated
on a hypermedia system called Microcosm at the University
of Southampton. Many of the principles used in this system
informed the Web as we know it today.

In 1987, Wendy and her colleague Gillian Lovegrove wrote the
paper 'Where have all the girls gone?' to address the deficit of
women working with computers. She continues to work at the
forefront of research in multimedia and hypermedia. She is a
Professor of Computer Science, Associate Vice President of
International Engagement, and Executive Director of the Web
Science Institute at the University of Southampton, and Managing
Director of the Web Science Trust to name but a few roles that she
currently undertakes. In these roles, she works tirelessly to ensure
that the Web and the Internet are governed responsibly.

As the Web has become more intertwined with AI technologies,
Wendy has brought her expertise to AI, and this is how we first
met. Wendy was writing a review for the British government on
AI. It was an exciting time, the whole community was abuzz, and
there was an obvious need to get more people with artificial
intelligence skills into the workforce. The many men and women
consulted on ways to achieve this suggested that graduates
who had completed undergraduate STEM degrees should be
the ones encouraged to do a Master's conversion course to AI.
When I first heard this idea, I was uncomfortable. I'd not taken a
STEM degree, and I knew that on average less than 25 per cent

of nationwide STEM degrees were taken by women. I shared my concern with Wendy that this project might reduce the number of women who'd be able to take up this opportunity. Despite her own maths degree and computer science prowess, Wendy had in fact already started campaigning to open up this conversion course to people from other subjects. Since then, through the AI Sector Deal and the Office for AI she's been able to get funding for non-STEM conversion, meaning that graduates from disciplines such as history, English and philosophy can access the course – richly diversifying the field of AI. There are more details about how you can apply in Chapter Eight.

Wendy has shattered many glass ceilings and has fought relentlessly to bring other women with her.

The 2000s

Fei-Fei Li (1976–) & ImageNet

Fei-Fei Li worked at the intersection of computer science and neuroscience for years before founding ImageNet in 2006. ImageNet is an enormous database of labelled images used for object recognition. It was set up by Li because she recognised AI needed a database as well as algorithms and formulae. Building ImageNet through meetings with Christiane Fellbaum – one of the leaders of WordNet doing the same for language – it is now considered absolutely invaluable for AI research. ImageNet led to the deep learning revolution (the period where deep learning became the standard in AI) by tweaking training algorithms to make them perform much better. Essentially, it changed the AI game.

What I admire so much about Fei-Fei Li is the way she has always been invested in making AI more accessible and diverse, as well as considering its impact. In 2011, she started the non-profit AI4ALL, which runs summer programs in AI for women, people of colour and low-income high-school kids.

Watson IBM Jeopardy

Remember IBM's Deep Blue chess match? Well in 2010, IBM developed a machine that was trained by AI to play the game Jeopardy, learning the right and wrong answers from a bank of old Jeopardy questions. They named it Watson. The machine had no access to the Internet, so couldn't search the answers online. Instead, it had to figure out the answers from what it already knew. It was part of a research plan to make the public more interested in AI and its potential. In 2011, Watson played two human contestants and won the game.

Voice Assistants

Another big moment in the use of everyday AI was the introduction of voice assistants. Siri was the first virtual AI assistant that could not only combine speech recognition and information retrieval but could also live conveniently in our pockets. Apple acquired it from its namesake company in 2010, then launched it as part of the iPhone the following year. In 2015, Amazon brought in Alexa, a device that responds to questions in real time, like Siri, but is activated by the sound of its name, rather than the push of a button. It was initially marketed as a way to make life easier: scheduling a calendar, playing music, connecting to other useful apps. But now Alexa can connect to smart home devices to switch lights on and off, and you might even hear her manipulated to rap on records.

DeepMind, Alibaba, Tencent, Baidu & OpenAI

DeepMind is a start-up AI research company founded in 2010 by Demis Hassabis, Shane Legg and Mustafa Suleyman in London. DeepMind is now run by a seriously awesome woman, Lila Ibrahim, and is a subsidiary of Alphabet, the most cash-rich company in the world (it even owns Google!). Like the chess games of the 1990s, the DeepMind team have been applying and developing AI techniques to the game Go – an ancient Chinese two-player boardgame, established over 3,000 years ago. In their computerised version, named AlphaGo, the number of possible board configurations is, to quote the DeepMind website, 'More than the number of atoms in the known universe.' By employing a whole range of AI techniques including reinforcement learning, AlphaGo won its first match against a human in 2015 before going on to win the Go summit in 2017. One of the wildest things is that after that 2017 summit, DeepMind released AlphaZero, which unlike IBM Watson was never taught from a trained dataset. Instead, it learned how to play from trial and error against itself, starting with totally random moves. This creates a way to win that human beings almost never do themselves. The DeepMind team see this as a critical step towards AI-augmented human creativity, and the potential of Artificial General Intelligence. Demis Hassabis is recognised as one of the biggest brains of his generation, so watch this space . . .

In the same year that AlphaZero was released, the Chinese government started pouring millions of dollars into AI development. Their focus was on helping organisations, companies and the military adapt and deploy AI, as well as integrating AI training into education. The Chinese government has plans to have built a $1 trillion AI industry by 2030. There are three major tech companies in China leading this AI revolution: Alibaba, Tencent and Baidu. These companies are described by journalists and governments alike as rivals to the American giants Amazon and Google. And what they're building definitely challenges the idea that the US is 'winning the AI race'. In fact, a recent book by Kai-Fu Lee called *AI Superpowers* examines how the USA and China are developing AI, and argues that China has the upper hand. For example, the Chinese company Alibaba is currently the world's largest e-commerce marketplace and has invested in seven AI research labs that will focus on machine learning, network security and natural language processing. And in 2018, Alibaba released robots that outperformed humans on the Stanford University reading test.

OpenAI

OpenAI is the AI research company founded by, among others, Elon Musk. In February of 2019, OpenAI created language software that was so good at learning human speech patterns and then generating text that they deemed it too risky to release the code to the public. This is because it could be misused, particularly for fake news and spamming people. But it's important to note that currently machines can't write perfect prose, nor can they have a true understanding of the world in order to be manipulative. What we need to be conscious of is that the people who are controlling them can't either.

In August 2019, OpenAI decided to release a more curtailed version of the modeller that you can actually play around with yourself without having to use code. Try entering some text and seeing what the OpenAI bot refers back with by visiting **https://talktotransformer.com**, entering a prompt that the AI will complete.

I fed the machine Spice Girls lyrics, and below is the result. I'll leave it to you to decide if we should have AI Spice on the reunion tour.

> **Prompt:** '2 Become 1', by the Spice Girls
> **Completion:** Now come in here and show me how it's done Now I'm watching a little history By the end of the night, I'll know what's going on And that, all I need are some deep thoughts I need some love.

When reading the history of AI, it's important that we consider that the ideas of 'winning', 'leading' and 'progressing' in AI are not synonymous with 'better'. We'll explore more about this in Chapter Four, but good questions to keep in mind as we're looking at more modern advancements are: what does it mean to be winning this race? What would you prioritise as a measure of success? What is the prize? And who gets to claim it and has been everybody been acknowledged? And even: who gets to decide what it means to be a leader of AI?

3

HOW DOES AI WORK?

Okay, so some of you might feel like skipping over this section and I get you: this stuff can be tricky to wrap your head around and maybe you feel like it's not necessary to know the ins-and-outs of AI. However, I know I always feel much more comfortable and confident with a concept when I take the time to try to understand its foundations. I believe it's worth doing the same here with the framework of AI, so that you can then extrapolate what you need when confronted with an AI system.

Karen Hao (my friend from the MIT technology magazine) and I discussed the basics of how AI works, and we hope that the following will provide a good springboard from which you can keep learning.

Let's take a look at how AI machines are trained, what we need to know about machine learning and the difference between it and AI.

Machine learning is a subset of AI algorithms that excel at finding patterns in data. On Netflix, for example, a machine-learning algorithm figures out what TV series and films you like (this is the pattern) based on what you watch, how long you watch it and dozens – even hundreds – of other factors that you don't even realise (this is the data). It then uses that pattern to recommend more programmes that you will probably like.

On Instagram, another machine-learning algorithm figures out what products and services you might pay for (the pattern) based on what accounts you follow, what you double tap and what you spend extra time on before continuing to scroll (the data). It then also uses that pattern to push you those targeted ads that seem to know you so well.

As you learned in the previous chapter, AI researchers spent many decades experimenting with various ways of replicating human intelligence. Machine learning proved to be the most successful. So successful in fact that almost every major app and website that you interact with now has commercialised it in some way to optimise, curate and tailor your experience based on your behaviours and preferences.

My first 'woah, AI is everywhere' moment was when I realised I no longer needed to tag my friends in pictures on Facebook because they were automatically tagged. Facebook identifies my friends and family from the other 2.45 billion people on the site – but how?
Facebook's photo-tagging feature is no different from the other apps and websites I've explained above. Every time you upload another photo of yourself or your friends to the platform, its machine-learning algorithms use it to teach itself the differences between your faces. Initially, it won't be very good – that's why you sometimes have to tag people manually or Facebook gives the wrong tag recommendations. But over time, as it sees more and more examples of your and your friends' faces, it

will get better and better at figuring out which group of pixels correspond to each of you. This is otherwise known as the training process. A machine-learning algorithm must 'train', aka learn the patterns within a dataset, before it can be applied in a new environment with new data.

Got it! What about other types of AI such as deep learning?

Just as machine learning is a particularly popular subset of AI, deep learning is also a popular subset of machine learning. Deep learning is like machine learning on steroids – it thrives at finding patterns in especially vast quantities of data, like billions, even trillions, of data points. This is the scale at which social media and streaming platforms operate. So really, Facebook's photo-tagging algorithm is a deep-learning algorithm. Netflix's recommendation algorithm is also deep learning: in 2018, the platform's users generated roughly 1 trillion data points *per day* that its algorithms then had to sort through and make sense of. To put this number in context: it would take 32,000 years for a trillion seconds to pass.

If AI can 'see', 'hear', 'speak' and predict things about the future, how can we better understand the technical elements of these skills?

We now know the template for machine learning: you feed an algorithm a bunch of data, it finds the right pattern in that data, then it uses that pattern to make decisions. We also know a little about how the algorithms of Facebook, Instagram and Netflix work following that template. So, before you read on, let's try an exercise: for each of the five categories above, make your best guess for **a)** what kind of data the algorithm needs, **b)** what patterns it looks for and **c)** what actions it takes.

As an example, here's what I would write for **Facebook's photo-tagging algorithm**:

- The data: the tagged photos that people upload
- The pattern: the groups of pixels that correspond to each person's face
- The action: automatically tag people in newly uploaded photos

Now you try.

Amazon Alexa's speech-recognition algorithm
- The data:
- The pattern:
- The action:

Gmail's Smart Compose tool
- The data:
- The pattern:
- The action:

YouTube's newsfeed-ranking system
- The data:
- The pattern:
- The action:

Spotify's recommendation algorithm
- The data:
- The pattern:
- The action:

Now let's work through it together following our template.

Image recognition

Facebook's photo-tagging algorithm is an example of an image-recognition system. For any such system, both the images and the image labels constitute the training data, and on Facebook, as we've talked about, those labels are your tags. Similarly, if you want to build an algorithm to recognise different dog breeds, plants species, or even someone's handwriting, you need to label or tag all your dog, plant and handwriting data. This idea of using labelled data to train a machine-learning algorithm is known specifically as supervised machine learning. You, in other words, are a supervisor, instructing the machine on what it needs to learn.

As an image-recognition algorithm trains, it strengthens its associations between which clusters of pixels correspond to which labels. Those are the patterns it learns, which it can then use to recognise things in images that it has never seen before. Here's an important distinction, though. While the algorithm can recognise things in new images, that doesn't mean it can recognise new things. It will only be able to identify the things present in its training data. So if you feed the algorithm only labelled images of golden retrievers, don't expect it to recognise a Rottweiler!

In summary:
- The data: labelled images
- The pattern: the clusters of pixels that correspond to each label
- The action: the labelling of images it has never seen before

Speech recognition

Amazon Alexa uses speech recognition, which, like image recognition, uses supervised learning. But the data is audio rather than images, and the labels are the words being spoken rather than the people and objects being shown. The algorithm follows the same training process: it trains on the labelled audio and learns to associate which sounds correspond to which words. The trained algorithm, also known as a **machine-learning model**, can then transcribe the words it has learned when listening to a completely new audio track.

This is how Alexa knows what to do when you ask it to set a timer or play your favourite song. In the background, its speech-recognition algorithm has been trained on millions of labelled audio clips of people saying various words. But like image recognition, speech recognition is also only as good as its training data. Alexa can't understand Mandarin Chinese, for example, because it hasn't been trained with any audio clips of the language.

Here is our summary:
- The data: labelled audio
- The pattern: the sounds that correspond to each word
- The action: the transcription of words from audio it has never heard before

Text Prediction

Gmail's Smart Compose tool is an example of a text-prediction algorithm. Text prediction is a little different. It uses what's known as unsupervised learning. In unsupervised learning, you're no

longer labelling any training data. The algorithm learns to find
the patterns itself. In other words, when you feed text into a text-
prediction algorithm, it learns which letters most often appear
next to other letters, or which words most often appear next to
other words. It then uses those associations to predict the letters
and words that might be used to complete a prompt.

This is why Gmail Smart Compose seems uncannily good at
completing your sentences. It's also what powers autocorrect on
your phone, and Google's and YouTube's suggestions when you
start typing your query in their search bars.

Unsupervised learning can still be written in our same template
for machine learning:

- The data: (unlabelled) text
- The pattern: the letters and words that most often appear next
  to one another
- The action: the automatic completion of words or sentences
  given a prompt

Ranking Systems

YouTube's newsfeed is ordered by a ranking algorithm, and how
ranking systems use machine learning might not be so obvious.
Like Facebook, Instagram and Google, it orders things in your
feed or search results based partly on what it thinks you want to
see first. The pattern these algorithms are trying to find is traits
about different pieces of content that might correlate strongly
with what you prefer. And the data these companies use could
be anything you've done on their platform that indicates some
kind of preference. Liking a post on Facebook or Instagram,
for example, is a pretty good signal to those platforms that you

enjoyed that post. Clicking on a particular video in your YouTube search results is another good signal that you'd prefer to watch it over the others.

Once a ranking-system algorithm has properly developed a model of your preferences, it can push content up or down the ranking accordingly.

To recap:
- The data: anything you do that sends a signal about your preferences (liking, clicking or otherwise engaging with a piece of content)
- The pattern: the characteristics in different pieces of content that correlate with those preferences
- The action: the ordering of new content based on those predicted characteristics

Recommender systems

Recommender systems, like Spotify's algorithm, are almost identical to ranking systems, except for the final action. The algorithms are still trying to develop a model of your preferences, but ultimately use that to recommend instead of rank the content you might like. That's what Spotify uses to recommend you new songs and what Netflix uses to recommend new programmes. TikTok, meanwhile, uses a recommendation algorithm to keep you watching and swiping for as long as possible. Adverts that target you are also a huge application of these algorithms. Google, Amazon, Instagram and any other site that seems to know exactly which products you would buy, has some version of an ad-recommendation engine running in the background.

To recap:
- The data: anything you do that sends a signal about your preferences
- The pattern: the characteristics in different pieces of content that correlate with those preferences
- The action: the recommendation of new content based on those predicted characteristics

To tie all this back to the bigger picture, deep learning – and machine learning, more broadly – are only one category of methods that researchers have tried using to mimic human intelligence. While such algorithms have had huge commercial successes, they're still a pretty shallow copy of our capabilities as humans and a long way off from anything that we would consider human-like. That being said, this one simple category has undoubtedly already had an immense impact on many aspects of our lives.

What AI Can't Do

It might seem like AI is magic, but it definitely has limitations, and we must remember that it is in the early stages of its development. It's a lot easier – and more exciting – to sensationalise and print the overnight success of a new tool than it is to publish the slow grind. That's why so often the press sensationalise a story, and why we need to have a critical lens when we're reading about what AI has managed to do.

The other times when AI makes the papers is with outrageous headlines like 'Terminator-style AI is only one to two decades away.' These can fuel hype and create unnecessary fear. But of course, the science is slightly more sophisticated. I'm going to break down one such headline for you.

'Facebook's Artificial Intelligence Robots Shut Down After They Start Talking to Each Other in Their Own Language.'

Facebook's experiment was to try and get two English-speaking chatbots to negotiate with each other and trade hats, balls and books. Each object had been given a different value, so the hope was that the bots would find a way to simulate people bartering in a flea market. What actually happened is that the bots figured out which keywords or phrases were most likely to result in the trade they wanted, and they started repeating those words over and over again. For example, 'Balls have zero to me to me to me to me' repeatedly. From the transcript, it looked like the bots were communicating and having a negotiation between them that humans couldn't follow and were therefore outsmarting the experiment. But the researchers knew this wasn't entirely the case: this experiment was in fact a failure. The whole point was to get bots to negotiate with people. While they may have found out how to make trades, they were using words in a way we do not – and nor did they create a language of their own. This is why the project was shut down. It was not that the bots had become some sort of terrifying communicators, scaring the researchers: they were simply nonsensical.

So whenever you hear about a supposed success or dystopian failure ask yourself:

- Does it sound too good to be true? Or perhaps like your worst nightmare? If the answer to either of these questions is yes, don't take the reported event at face value: look into it.
- What data was it trained on and who is it representative of?
- How have humans been manually completing these tasks before and is this AI system really better at the whole process – or only one aspect?
- Who paid for the research or product? And who stands to benefit?

Most importantly, there are parts of our lives in which AI is not capable of playing a role, nor necessarily would we want it to. AI can't forgive, care, relate, show empathy or have morality – more on this later.

4

HOW NOT TO TALK TO ROBOTS:

The Potential Risks

The potential of AI is enormous – no one can dispute this. PWC, a huge consulting company, predicts AI could add 15 trillion dollars to the economy by 2030. To put that figure into perspective, for the next fifty years you could spend $41 billion every day and still have some money left over.

It's precisely because AI has so much potential that it presents both life-changing rewards *and* potentially catastrophic risks. This is why I believe it's so important to wrap our heads around its complexities and for you to have information to start weighing up the pros and cons.

Since the list of ways AI may affect us is pretty much endless, I've chosen a few of the most thought-provoking issues to spark your imagination. I hope that they start you on your journey looking into the potential of AI; and using the reading list at the back, you can read more about the areas that worry or excite you the most.

What Are the Risks?

The media can sometimes be a fearmonger, using dramatic headlines that fight for your attention, which is – thanks to fancy algorithms – worth millions to them. This has created a wealth of material concerning the potential AI has to wreak havoc on our lives. When you start reading or listening to these arguments, they often sound compelling or extremely alarming, or both at the same time. My aim here is *not* to generate fear about AI, but instead draw attention to what could happen if it goes unchecked or is left in the hands of the few. My hope is that if the risks are brought to our attention now, we can use that information to make more informed decisions about such things as the products that we use, media we consume and companies that we work with or for. I've chosen examples that are current at the time of writing and as relevant as possible.

Risk 1. The AI is programmed to create and spread dangerous content

From the beginning of time, media and social interaction within a community has been leveraged by brands and politicians to capture us as customers and win our votes. The dawn of the Web enabled this sort of propaganda to spread further and wider and AI takes the lid off: for not only can AI be built to generate content, it can be used to target people and disseminate topics more efficiently and effectively than ever before. Here are two strikingly different but equally potent real examples associated with this technology.

Let's start with the most famous case of this, that of Cambridge Analytica and the 2016 Brexit Referendum. Cambridge Analytica were alleged to have shared Facebook data with the political party UKIP in order to identify people in the UK who were likely to be 'persuadable' in whether they voted to leave or remain a part of the European Union. UKIP were said to have then targeted this group of people by dropping curated content into their social media feeds with the intention of tipping their neutrality into a Leave vote. Those targeted were unaware that their personal data had been used for this purpose or, indeed, that they were the victims of targeted ads. Both Cambridge Analytica and UKIP deny these claims. Crucially, none of us would even know that this had happened if it wasn't for a *human* whistle-blower. If you want to learn more about Cambridge Analytica and how our data is sold, I'd really recommend watching the documentary *The Great Hack* or reading Christopher Wylie's book *Mindf*ck*.

Sometimes the risks of AI can feel even more deceptive. This is true with the rise of what's known as deepfakes – videos that harness the power of AI to blend existing video content with fake content to create fake videos that are so shockingly similar to real ones it's hard to tell they've been doctored. Currently, 97 per cent of deepfakes are being used to make pornography, with famous actresses almost always the target of this subversion, although increasingly women who are not famous are also being targeted. There is no need for there to have been real pornographic content in order to make a porn deepfake. What's more, it's becoming easier for anyone to make these videos; you don't have to have particularly expensive technology to hand. All you have to do is upload a video and use a program to type what you want the person in it to 'say'. This is deeply troubling, and as I write this book, strides have been made in detecting and then legislating against deepfake videos. But since we are capable of creating a world where we can't rely on what looks real to actually be real, vigilance is always going to be our greatest weapon when we're looking at any type of content online.

Risk 2. The AI is programmed to do something beneficial but physically harms or discriminates against certain groups of people

Sometimes the harms of AI come from deliberate malicious programming, but many of the serious consequences of AI are unintentional and stem from programming that has unconscious bias coded into its design that wasn't caught before it reached the user.

Joanna Bryson, Professor of Ethics and Technology at the Hertie School, produced a paper when at the University of Bath that, just from looking at words and how words are used, showed how you get an enormous amount of human experience and how this can be fed into the AI. The paper showed that, on average, men are associated with work, maths and science, and women with family and the arts. And young people are generally considered more pleasant than old people. She explains that algorithms make decisions based on which words appear near each other frequently. If the source documents reflect gender bias – for example, if they more often have the word 'doctor' near the word 'he' than near 'she', and the word 'nurse' more commonly near 'she' than 'he' – then the algorithm learns those biases too. They called this 'word embeddings', basically a computer's definition of a word, based on the contexts in which the word usually appears. So 'ice' and 'steam' have similar embeddings, because both often appear within a few words of 'water' and rarely with, say, 'fashion'.

But to a computer an embedding is represented as a string of numbers, not a definition that humans can intuitively understand.

The gendered risks of a world built on data that has a 'default male' has serious consequences. This can be referred to as the gender data gap, and it's everywhere. Often the potential harm to women is physical, and in some cases can prove deadly. For example, when testing for car safety, the default test dummy used is the same as the average male in height, weight and musculature. This means that most of the data collected about car safety was made for men's bodies. Studies show that while men may be more likely to be involved in a crash, women are 47 per cent more likely to be seriously injured in one than a man, and 17 per cent more likely to die. It wasn't until 2011 that women's test dummies were introduced in the USA, while in the UK there is still no required safety test with a woman in the driver's seat (although one exists for the passenger's seat).

It's not only a gender data gap. Those who are historically underrepresented in data sets are also underrepresented in the design of AI.

Joy Buolamwini, an MIT researcher who you'll read more about later, has become famous for calling out this sort of bias, particularly in facial recognition. She conducted a study at MIT that found that Amazon's Rekognition face-recognition software perfectly classified the gender of lighter skinned men, but overall misclassified 19 per cent of all women and 31 per cent of dark-skinned women were misclassified as men.

This is about training data and who is used as the representative sample. The machine can only learn what's inside the dataset it's been given. It follows that if the dataset is primarily white men, it's going to be really good at classifying white men. If the researchers then test the algorithm on white men, it's going to look like a success. Once the algorithm is later used on a diverse dataset however, the system will have trouble recognising women and women of colour in particular.

The consequences of this can be far reaching. Facial recognition technology has already been implemented at border controls and is also used in policing – both areas where racial discrimination is rife.

The key thing to understand here is that AI is not itself biased. It reflects the data it's trained and the world in which it learns. A good example of a machine's software unintentionally reflecting a misogynistic and racist 'upbringing' is Tay – Microsoft's AI-powered bot. Tay was built to perform a number of tasks, such as telling jokes or commenting on photos posted to her Twitter account. She was also designed to adapt to her environment. Although Tay was not coded to be racist, she proceeded to learn and evolve personality from the users online. When users made racist comments, Tay would echo them back in its own commentary. Tay's tweets began to include racial slurs, references to Hitler, denying the Holocaust and support for Donald Trump's immigration plans. All these happened within twenty-four hours of her release on Twitter in 2016. Microsoft switched off Tay when its behaviour began to cause extreme offence. Yet there is still technology in use that is less visible, less blatantly offensive or discriminatory, and therefore almost immune to public backlash. These continue to thrive in the shadows and so it's worth remembering Tay when you consider how a new product is being trained.

'AI, Ain't I a Woman?' by Joy Buolamwini

My heart smiles as I bask in their legacies
Knowing their lives have altered many destinies
In her eyes, I see my mother's poise
In her face, I glimpse my auntie's grace
In this case of deja vu
A 19th century question comes into view
In a time, when Sojourner Truth asked
'Ain't I a woman?'

Today, we pose this question to new powers
Making bets on artificial intelligence, hope towers
The Amazonians peek through
Windows blocking Deep Blues
As Faces increment scars
Old burns, new urns
Collecting data chronicling our past
Often forgetting to deal with
Gender race and class, again I ask
'Ain't I a woman?'

Face by face the answers seem uncertain
Young and old, proud icons are dismissed
Can machines ever see my queens as I view them?
Can machines ever see our grandmothers, as we knew them?

Ida B. Wells, data science pioneer
Hanging facts, stacking stats on the lynching of humanity
Teaching truths hidden in data
Each entry and omission, a person worthy of respect

Shirley Chisholm, unbought and unbossed
The first black congresswoman
But not the first to be misunderstood by machines
Well-versed in data drive mistakes

Michelle Obama, unabashed and unafraid
To wear her crown of history
Yet her crown seems a mystery
To systems unsure of her hair
A wig, a bouffant, a toupee?
Maybe not
Are there no words for our braids and our locks?

Does sunny skin and relaxed hair
Make Oprah the first lady?
Even for her face well-known
Some algorithms fault her
Echoing sentiments that strong women are men

We laugh celebrating the successes
Of our sisters with Serena smiles
No label is worthy of our beauty.

This poem was written by **Joy Buolamwini**. She describes
herself as 'a poet of code' – a term that I adore for its
expression of how art and technology flourish when
combined. She also coined the phrase 'The Coded Gaze',
which will resonate for those of you who have read about
the 'Male Gaze' in 70s feminist film theory.

We discussed including her poem in this book, as I often hold these words in my head when I look at news headlines about breakthrough technology and now you can too. Online, the poem is accompanied by a video that you should watch. It shows AI failing to register the faces of iconic black women like Oprah Winfrey, Michelle Obama and Serena Williams, serving as a stark reminder of how racism creeps into technology.

She uses her research and art to illuminate the social implications of AI, founding the Algorithmic Justice League to create a world with more ethical and inclusive technology. Joy launched the Safe Face Pledge, the first agreement of its kind that prohibits the lethal application of facial analysis and recognition technology.

Risk 3. The AI is programmed to do something beneficial, but creates a dependent relationship on machines

Sometimes, the risk isn't about the technological impacts themselves, it's the way it changes our behaviour with that technology – or the way its existence changes the way humans interact with each other. You'll see that Kate Devlin and Beth Singler both talk about this in their interviews in Chapter Five. An example we can look at now concerns Internet addiction and AI.

A lot of AI algorithms, especially those that recommend things you might like, have been built not only to help us find what we want, or what their maker thinks we want, but also to be deeply addictive and keep us logged on. In fact, many studies draw links between how AI is designed and slot machines. The Mozilla foundation describes this as 'addiction by design'. Further to wasting time, prolonged use of the Web can have a profoundly negative affect on our health causing dependency and isolation issues.

The impacts aren't only that it wastes time and is isolating. YouTube is an example of how the benign desire to keep us online can have a catastrophic impact on how we see the world. YouTube didn't always use a recommendation algorithm as part of its platform. But now when a user is watching a video, there are a whole bunch of features on the site that use recommendations to try and keep the user watching. Take the auto-play function: it's personalised based on what the site knows you've already clicked on. This might seem great, because it's giving you more of what you want. But the problem is that it's been widely reported that the algorithm has figured out that the most clickable links (which is coded as success for the algorithm) tend to include extreme statements and even hate speech. So while it may appear that you're simply using the Web to discover an Ikea hack, or how to bake the best apple pie, remember that the relationship being built in those moments between human and machine is complicated.

Risk 4. The AI is programmed to do something beneficial, but automate part of someone's job and they risk losing it all together

AI is becoming ubiquitous in the workplace. From 2010 to 2018, the number of start-ups using AI increased eightfold. A recent study looking at the use of AI in nearly 3,000 enterprises worldwide estimated that now 14 per cent use AI. It's thought that by 2025, 80 per cent of enterprises will have implemented some sort of AI strategy into their portfolio.

Just because AI is often tied to automation this doesn't mean that everyone will lose their jobs. Tech entrepreneur Kathryn Parsons has some great insights about this in her interview. Automation will, however, effect the kind of jobs available to us. McKinsey Global Institute (a huge management consultant company) estimate that 60% of occupations have a 30% possibility for automation based on what AI can currently do. What's worrying for women, is that although it also depends on education level, age and nationality, in general, women's jobs are more likely to be automated. Statistically, there is an 11% risk of women losing their jobs versus a 9% risk for men. Women tend to work clerical and easily automated jobs, and they also make up only an estimated 22% of the AI workforce, which is one of the fastest growing industries.

Risk 5. The AI is programmed to do something beneficial, but infringes on our privacy

As AI technology grows in power and capability, so too does its ability to understand and classify human beings. But there are risks involved with AI's growing capacity to survey human beings. For example, researchers have found a shortcut to diagnose using AI whether a young person is experiencing mental illness. On the surface this sounds great and important and necessary for helping young people. But it comes at a cost, and this time it's the price of data privacy. The questions we need to ask here are: who controls the data? Who receives the information if someone is deemed by the AI to be 'at risk'? Will that person be fully trained to help that young person? Does the algorithm get it wrong, either with a false positive, or false negative, diagnosis? Carly Kind, director of the Ada Lovelace Institute, talks about this in relation to something named 'mission creep', which refers to the drift between the original intention of technology and the areas of our lives it may impact in the future. An AI that predicted you had mental health conditions while at high school may eventually be helping make decisions about where you can attend college.

Selfie apps are another great example of how people aren't aware of the rights around their images. One popular photo-storage app was recently reported for using its users' photos to train facial recognition software that it then sold to law enforcement. IBM was also found to be using Flickr photos to train facial recognition applications without explicit permission from those in the photos. PopSugar's viral 'Twinning' app inadvertently leaked data. FaceApp, who go viral with their latest technologies, did not have the most watertight privacy policy either. No one seemed to delve into this because the filters were so fun – and that was prioritised over privacy. In addition to photos generated via the app, FaceApp's privacy policy states that it also collects information about location and about users' browsing history. 'These tools collect information sent by your device or our Service, including the web pages you visit, add-ons and other information that assists us in improving the Service', the policy says. And though it states, 'we will not rent or sell your information to third parties outside FaceApp', it explicitly says that it shares information with 'third-party advertising partners', in order to deliver targeted ads.

It's often reported that smart devices like our phones are 'listening' to us when we haven't instructed them to. However, this is an urban myth and has been debunked and disproven. Researchers from the cyber security sector recently conducted a series of experiments looking at data usage. They determined that the phones tested were not 'listening' to the sounds in the test room. Any relationship between what people say and what they get targeted ads for is likely the result of these companies having collected huge amounts of data about us, and their using extraordinarily accurate AI-driven prediction and recommendation algorithms. I'm not sure which is more concerning, a phone and therefore advertisers listening, or the fact that the advertisers don't need to listen to predict what we want! Either way our privacy feels infringed. The key thing to consider here is that it's not your individual data that's powerful; it's what happens when your information is put alongside other people's in order to train an AI system about people like you. This means that your privacy can be compromised because the system surmises how you'll react, rather than having the information. You need to be aware of this when it comes to technologies that might be introduced in the future. For example, some studies found that certain apps sent screenshots of what users were doing on their phones to advertisers. Until there is stronger regulation put in place, we can't foresee the lengths advertisers will go to win you as a customer.

When it comes to privacy there is a longstanding debate over how to balance individual privacy with public safety. The tech giants are working to implement end-to-end encryption across their messaging services. This is the highest form of security that doesn't even allow for lawful access to the content of communications. The UK and US and other western governments have been campaigning for a 'backdoor' into these technologies. They cite a need for access to citizens' private messages to facilitate investigations related to terrorism, child abuse, exploitation and other serious crimes. While this may sound reasonable, it also shows the lack of understanding in the technology as Neema Singh Guliani, the senior legislative counsel for the ACLU explains, 'When a door opens for the United States, Australia, or Britain, it also opens for North Korea, Iran, and hackers that want to steal our information.'

Risk 6. The AI is programmed to do something beneficial, but makes us sick

AI is increasingly being used in the healthcare sector for a range of beneficial research and you'll hear more about this in the rewards section. The danger of AI being used in this way is that if it isn't trained on representative data, it can make certain people sicker. This can have even more serious consequences when used in diagnoses. For example, the symptoms of having a heart attack are subtle, and if you trained an AI only to recognise male symptoms, you'd have half the population dying from heart attacks unnecessarily.

Another horrifying example of askew health data specifically affects women of colour and their level of maternal healthcare. At the moment, many companies are trying to develop ways to use AI to improve the prediction rate of conditions like hypertension and preeclampsia, or others that can cause serious complications during pregnancy. On the surface, this seems like it could only be a good thing, but so often medical data is not representative of *all* women, nor does the healthcare system treat everybody equally. For example, in America, black women already have a much higher rate of maternal mortality than white women (regardless of education level or socio-economic status) as a direct result of systemic and institutional racism. So when AIs are trained to make decisions using skewed data, their results reflect and therefore perpetuate existing bias (including racism, prejudice and discrimination), and when it comes to healthcare, this can have direct life or death consequences.

Risk 7. The AI claims to protect us but creates new harms to wellbeing

Another example relates to working conditions. One thing that all big social media platforms need to operate safely is content moderation. This means they need AI tools to quickly identify and flag images and texts that might be harmful, violent or obscene. AI has proven to be incredibly successful at this task. Last year, Pinterest reported that they were able to decrease self-harm content by 88 per cent once they had implemented AI techniques. Facebook had a similar statistic. This may sound like a win-win scenario but as we've learned, nothing 'automated' is entirely automated. Machine learning works by recognising patterns and sorting accordingly. To classify an image as harmful or not, you need a big bank of images that teach the system what harmful content looks like. The first pass at this is completed by humans. What that means, is that in order for users to be protected from potentially harmful content, someone somewhere else is repeatedly exposed to difficult and challenging images. This work, usually referred to as content moderation, always hides behind these systems. It has been widely reported to be high stress, traumatic and low paid. One investigation conducted by Verge.com tells the story of a young woman named Chloe who spent every workday watching an average of 400 clips a day of extremely violent videos, including those depicting murders. What's more, most of this kind of work is outsourced to the Global South where labour is cheaper. Automated content moderation is extremely necessary for the well-being of users, it's just that it's currently at the expense of the workers' well-being.

Content moderation is not necessarily always for people's benefit, either. In some countries, the leadership uses AI to track and oppress voices with which they don't agree. Speaking out against the state can be monitored online and people caught out for dissenting. This may sound dystopian and far removed from our own reality, but it is real and it's happening across the world. I realised this when I watched a girl on TikTok doing a makeup tutorial. She was sharing her fear of the oppressive regime she was under and calling for help at the same time as curling her eyelashes. She knew exactly the behaviours and timing of a makeup tutorial and followed them immaculately as she interspersed gestures with political messaging so as to avoid being picked up by the AI system. It was beautiful and courageous all at the same time.

Risk 8. The AI becomes more intelligent than humans

Okay, so this is the biggie that many of the most vocal and powerful (men) in our society use their platform to lament. AGI (Artificial General Intelligence), also known as Super Intelligence, is often cited as something that could wipe out humanity. The presumed risk is that if we make super intelligent machines, then that would enable them to override their programmers and then destroy humans because their goals are such that humans are superfluous at best or at worst a threat. Famous names like Elon

Musk and Stephen Hawking, and other celebrity academics, wax lyrical about the terrifying consequences of AGI. It's been claimed that humans risk becoming 'pets or pet food.'

Kate Crawford is the co-director and co-founder, alongside Meredith Whittaker, of the AI Now Institute at New York University. It's the world's first university institute dedicated to researching the social implications of artificial intelligence. Kate argues that the issue of AGI is also one about quality of life. As you'll have seen from this chapter, for many people AI is already seriously hindering their ability to lead a 'liveable life', to borrow a term from the feminist writer Judith Butler. As Kate wrote in the *New York Times* 'Currently the loudest voices debating the potential dangers of superintelligence are affluent white men, and, perhaps for them, the biggest threat is the rise of an artificially intelligent apex predator. But for those who already face marginalisation or bias, the threats are here.' I agree with Kate. It's much more important to consider how AI is built today rather than focusing on the existential risk.

It's also important to note that leading researchers can't agree on whether AGI is even possible. Most agree there are twelve fundamental technological hurdles to leap, which in isolation may take anything from five to forty-five years to develop. When polled on the day we'd reach AGI, many predicted that we are 150 years away, while some believe it could be the next fifty years, and a few suggest sooner. Ultimately no one can say for sure.

Demis Hassabis, DeepMind CEO and Founder, is my oracle in many ways about this because he is on a mission to what is described as 'solve intelligence.' In his view, public alarmism over AGI obscures the great potential of near-term benefits and is fundamentally misplaced, not least because of the timescale. 'We're still decades away from anything like human-level general intelligence,' he says. 'We're on the first rung of the ladder. We're playing games.'

Risk 9. The AI is programmed for war

As has been the case historically, the race for weapons has often been the driving force for huge technical innovations – the Second World War and the Space Race are just two examples we looked at in the chapter on history. But AI weaponry is a different beast, with different scales and possibilities for catastrophe. Some of these AI-based weapons are already being used: drones are one familiar example. But once running, automatic weapons select targets without any further intervention from human beings. We need to be asking: How do you stop it? Who gets to program it, and to what end?

Although there are some arguments that have been made for AI weapons – they keep humans off the battlefield is one that often comes up – many of the big players in AI are so worried about these issues that they've released statements or agreements not to develop or program autonomous offensive weapons. For example, after mass protests and walk-outs, Google banned the development of autonomous weapons in 2018 and refused to renew a contract with the USA's Department of Defense. This was part of a larger campaign called #TechWontBuildIt, and both Amazon and Google workers have used the hashtag. The refusal of workers to contribute to projects of war, or more recently ICE (Immigration and Customs Enforcement) in America, demonstrates the importance of bottom-up, and vocal, informed workers being present in all sectors of the tech industry who understand the threat AI poses and how best to regulate it.

Risk 10. Global competition leads to bad decisions

As with the nuclear race, there is a real risk that the countries with different ethics, different politics and different cultural norms advance AI in ways that aren't safe or considered appropriate in another country. This could cause other countries to cut corners and make difficult decisions about how to stay ahead in the AI race. A good example of how this might work is to consider the ways that different hubs imagine the Internet. An article by Wendy Hall and Kieron O'Hara in the *Financial Times* explains that there are at least four competing visions for the future of the Internet. Two of these exist in the United States. While Silicon Valley pushes for an Open Internet, where there are no restrictions on data flows, there's a contrasting vision held in Washington DC that the Internet should be commercialised. Outside of the States, the EU advocates an Internet that fosters freedom, while also recognising the necessity of regulation. In Beijing, the Internet is talked about as a servant of the public good. This means it is heavily regulated by controlling data flow. It's a complex topic, and many others have dedicated entire books to exploring it – check them out in the further reading section at the back of this book.

What Causes These Risks?

In order to combat any of these risks, it's important to understand the cause of them. We can't be sure what is causing each individual issue, but these are some of the agreed-upon concerns of the AI community, and it's really worth getting to grips with them.

There are many moments that led to the rise of AI. Here I want to look at six that have caused particular concerns. One is technical and the other five are about a lack of representation. Before we go on, I want to stress that these five are by no means distinct; rather they often enable, reinforce or even cause one another. It's this kind of cycle that makes AI dangerous, because the impacts are not only multiple but also in aggregate.

In other words, the same groups are likely to be impacted by risks because they are systemically excluded from all stages of AI development including initial data, or as coders, programmers and regulators. Together, this makes for a perfect soup of bias and innovation. Lots of people I've included in this book will talk about this but below I've collated some examples to get us started.

Cause 1. Biased Data Sets

As you read in Chapter One, the power of AI systems lies in their ability to analyse vast amounts of data. Using this information, it can find patterns and make super-quick links.

The advent of AI systems originally promised the elimination of human bias as humans, with their biases, were supposedly taken out of the loop. It was often claimed, for example, that using an algorithm for hiring would give men and women an equal chance in job opportunities. It was believed that a machine wouldn't be able to judge you based on the colour of your skin, gender or postcode. This has been proved wrong. A machine can only learn to make decisions based on the data that it's been given and so that data must always be interrogated.

We think of AI as being objective because whatever is happening beneath the hood of it is maths, and maths is numbers, and numbers can't be racist – can they? But what if the numbers and commands we're feeding the AI is biased? We already know that the data the machines are learning can be biased. Not only is the data set biased, but the person working with the data has their own bias. Even if it isn't a conscious one, it will add a second layer of bias to the final code. As Cathy O'Neil said in her TED Talk, 'Algorithms are opinions embedded in code.'

Cause 2. Lack of diversity in the workforce

Google the phrase 'man and machine' and your screen will fill with images of men shaking hands with robots. It's a scene that almost parodies Michelangelo's fresco *The Creation of Man*, extolling the symbiotic relationship between men and technology. But search for the phrase 'woman and machine' and you'll be bombarded by women provocatively caressing washing machines. Totally warped, right?

Sadly, the tech industry in the USA and UK is still dominated by men, and for decades its white, male-focused programming has pretty much gone unchecked; these search results are only reflecting the current status quo. This has a lot to do with how society raises women to believe they have a particular place in the world. Stereotypes about gender start young. By the age of three, we can already observe patterns of behaviour emerging that distinguish 'boy' from 'girl', and this is echoed in simple but actually quite scandalous ways – like how toy shops arrange aisles into pink toys for girls and blue toys for boys. Research from the University of Washington's Institute for Learning & Brain Sciences showed that by the age of six, children already believe boys are better than girls in programming and robotics. This shows the inherent sexism as this is of course neither technically nor physically true.

Gender bias doesn't happen in isolation, it's always intertwined with issues of race, class, sexuality and ability. As we've seen, many of the world's original computer programmers were women, but the marginalisation of women's presence in technology happened around the same time personal computing become a lucrative industry. Discrimination and sexual harassment have driven women out of Silicon Valley and kept countless more from entering. A 2017 WISE report showed that women hold only 23 per cent of STEM roles in the UK and a LivePerson study demonstrated that out of a random sample of 1,000 people in the USA, only 8 per cent could name a famous women in tech. Even when they could, half of those people named Siri or Alexa!

When it comes to AI, the statistics are even worse. Nesta, the British innovation foundation, published a large-scale analysis of gender diversity in AI research in 2019. The report found that just 13.8 per cent of the authors of AI technical research were women.

Cause 3. AI is a black box

Data scientists don't actually have access to what is underneath an AI system's interpretations of the world. This is often described as a 'black box'. Because the process is hidden, it's particularly difficult to identify if the data that feeds into a particular conclusion has been shaped by inequality, bias or discrimination. Access to data, due to the proprietary nature of AI systems, means interrogation is nearly impossible. By masking these sources of bias, an AI system could consolidate and deepen already systemic inequalities, simultaneously making them harder to observe and challenge. It's like driving a car without being able to open the bonnet. Sandra Wachter talks about this at length in the next chapter, explaining in more detail some of the ways the AI community is addressing the issue.

Cause 4. Lack of accountability

If you think about all these risks and then consider who or what is responsible when something goes wrong, it can get a bit complicated. The very nature of AI is that it's autonomous and hard to know why an action was taken. This is why the technologists building products have been known to blame the tech rather than themselves when the AI system behaves in ways that harm people.

Take a self-driving car for example. Even if the vehicle makes a completely autonomous decision to leave a motorway at high speed and crash (perhaps foreseeing an oncoming crash and preferring to take preventative action), you obviously cannot take that self-driving car to court to force it to face justice. Lawmakers are currently grappling with the challenge of who should be accountable for an accident like this. Who is the 'responsible driver' in such a scenario?

We know these AI algorithms are created using vast amounts of data, so where *should* the blame lie? With the people who collected the data about the roads? Or the engineer who wrote the maths that trained the AI to have vision? How about the manufacturer that rolled out the AI system in all their cars? Or the CEO who decided to follow an autonomous strategy? Does the blame lie with the driver?

From my perspective, the accountability needs to lie with the management, the people who took the decision to build the AI and then signed the system off ready for customers. Laws will need to change to hold businesses and their executives to account. Companies will need even stricter and more robust processes to ensure responsibility is felt across the organisation and therefore safeguards put in place.

The challenge we have with new technology, especially AI, is that it's difficult for regulators and governments to keep up with the pace of innovation. This is why we need to make sure that individuals understand the ramifications so that they are at least aware, even if not fully protected.

Cause 5. Digital Divide

Lack of diversity isn't only a problem in terms of who is working in the field, it's also about who can access and use the new technologies that are built with AI. Poverty and age are predictors of whether or not someone will be able to access digital technologies, an inequality referred to as the digital divide. And with growing rates of income inequality and an ageing population, it's an issue that urgently needs to be addressed.

Like so many of the causes I've mentioned here, the digital divide is a cyclical problem. Not only do these groups not benefit from new technologies, but it also means that there is no data collected about them or their needs. This prevents products from being built to improve their lives, and the chasm grows. The system reflects the existing injustices and widens the gap between the wealthy and the excluded populations. This has to be understood, and the challenge of addressing the gap needs to be accepted, before new products and services are developed to truly close the digital divide.

What is the AI community doing to avoid these issues?

The AI community know that the risks above cannot be left unchecked and there is no such thing as machine neutrality. Existing discrimination towards women, and especially women of colour, *will* be further exacerbated if inequality continues to be codified into our machines. For both the AI community, and individuals outside of the tech industry, vigilance is key going forward. There are, however, ways to counter machine bias and the most important thing is to demand increased transparency and accountability at all levels. There will be more on this later in Chapter Seven. Technology companies cannot keep how intelligent systems operate secret for commercial reasons. They have to be publicly available so that if something starts to go wrong, it can be spotted and put right. In Chapter Six, Wachter talks about this more, and in Chapter Seven, I'll be offering some practical advice on how to tackle these issues yourself rather than just waiting for regulation. First, let's explore more of what the AI community has already put in place.

Explainability Tools

Remember the black box problem from the risk above? Well there's a movement within the AI community that is trying to tackle this problem using explainability tools. These are methods to increase transparency so that the people working with algorithms have a better understanding of what actually happens during the process of a machine making a decision. This can then benefit the consumer because if you want to know why you were denied a loan, for example, or didn't get a job, then you could ask the organisation or company to show you what factors contributed to that decision.

Regulations like GDPR

In 2018, the GDPR came into effect in the UK. General Data Protection Regulation is a European privacy law governing data rights on the Internet. It was an important milestone because it replaced laws that were written for what the Internet looked like in 1995 – when I was first using Clippy to help me navigate Word. But what most of us don't realise is how much it can protect individual data rights, in addition to those of companies. Because so much has changed in how the Internet is used for business, and the role data and data-sharing have in our lives, the new GDPR increases restrictions on what organisations can do with your data, and it extends the rights of individuals to access and control data about themselves. In some cases, it also extends these restrictions and safeguards on what can and cannot be done with your personal data to organisations based outside the European Union if they handle data collected within it. The GDPR requires organisations handling personal data to adhere to its six data processing principles. You can look these up, but this

isn't a book for organisations – it's a book for you as individuals.
So I want to share how the actions the organisations must take
will affect how *you* live and work online. Too much of the roll-
out of GDPR ignored the fact that this is a win for you – and
there are means through which you can go some way towards
keeping your data-self safe. I say 'some way' because it's widely
acknowledged that GDPR is not enough to protect data-driven
decisions made using AI, and is being reviewed.

There are eight key rights to know about which you can invoke
in different ways, as discussed in the chapter on actions, Chapter
Eight:

1. Right to be informed: you must be told what data companies
    will collect about you

2. Right of access: you can now download all the data a
    company already has on you

3. Right to rectification: you can have incorrect personal data
    amended

4. Right to erasure/to be forgotten: you can have your data wiped

5. Right to restrict processing: you can limit how your data is
    shared and analysed

6. Right to data portability: you can duplicate and move your
    personal data

7. Right to object: you can limit what data is collected on you in
    some instances

8. Rights in relation to automated decision-making and profiles
– you have some special rights if a computer makes an
automated decision relating to you

Now that you know more about this regulation, think about which
additional rights and rules you believe should be included.

Rules, Standards, Codes of Practice

In addition to government inventions, there are also other bodies,
formal and informal, that are putting their heads together to
create new rules, standards and codes of practice for AI. Since
2017, this has begun to occur more frequently and with more
urgency. One example is a report published by the Royal Society
and the British Academy led by Ottoline Leyser and Genevra
Richardson, 'Data Management and Use: Governance in the 21st
century'. This report suggests there should be one overarching
principle to govern the intelligent machines we'll soon be living
alongside. The main idea is that 'humans should flourish', which
makes a lot of sense to me. This phrase rightly leaves lots to the
imagination and very much depends on individuals and societies
working out new understandings of what it means to flourish.
Claire Craig, one of the team involved, explained to me that they
wanted to get away from a solely Western, individualistic notion
of flourishing. The report proposes we centre people in the
design of AI, especially the idea that humans – as a collective –
must always be in control of these machines. It recommends four
high-level principles to promote human flourishing that I think
still form the basis for considering the human relationship with AI.
These are:

● Protect individual and collective rights and interests
● Ensure transparency, accountability and inclusivity

- Seek out good practices and learn from success and failure
- Enhance existing democratic governance

These kinds of reports have continued to be a force for good in the AI community. For example, places like Doteveryone work to make responsibility the norm in AI development across different sectors, while the Partnership on AI brings together leaders from the non-profit and private sectors to develop best practices for AI technologies. In addition to these, many individual companies have their own internal codes or standards to ensure the AI they build helps human beings rather than replacing them. You'll hear from the founder of Doteveryone, Martha Lane Fox, in Chapter Seven.

Human in the Loop

Human-in-the-Loop (HITL) is a way to build AI systems that makes sure there is always a human with a key role somewhere in the decision-making process. This kind of system guarantees that whatever the outcome happens to be, it's arrived through a combination of steps taken by a machine and the person, together. What I want to emphasise is that the system works as a loop, not in a straight line with a beginning and an end. In other words, the system is constantly getting feedback from the human, and the human from the system.

One simple way to illustrate HITL is in image classification, let's say between dogs and cats. A group of people will create the original labelled data, and a machine will use those labels to learn what cats and dogs look like. Then the machine uses what it has learned to group a whole new set of images into cats and dogs. For those the machine doesn't know what to do with, the human comes back into the system, labelling what the machine couldn't. Those trickier labels are then fed back into the machine so it can continue to improve its accuracy. There are plenty of other examples of how HITL works, and while it is helpful for tasks like the one described above, it is most critical for systems where if the machine makes a mistake, the consequences are deadly, such as with autopilots or autonomous vehicles. Hannah Fry explains more about this in the next chapter.

Diversity in the workforce

In the 'risks' part of this chapter, I listed a lack of diversity as one of the causes of biased and discriminatory AI. The AI community needs to encourage diverse voices to contribute to and lead the discussion about this technology because although AI will impact all of us, the way these impacts will be felt are different depending on our experiences and positions in society. Thankfully, parts of the AI community have started to implement ways to create a more representative workforce. For example, some companies have removed gender identifiers from CVs of candidates to encourage gender-neutrality in hiring. Also, many organisations provide free coding lessons to communities that are currently underrepresented in the field. These programmes give women, people of colour and those from low socioeconomic backgrounds access to skills that may have otherwise been denied to them, so they can enter the tech sector. In Chapter Five we discuss how Anne-Marie Imafidon's social enterprise 'Stemettes' does exactly this, while Jess Wade explains more about how tech is being used in the education system.

Open Source and Open Data

The principle is that the code and data that goes into the machine to make the AI work should be made available for everyone to access. This means that people outside of the company, organisation or research team would have permission to use and study the content. Not only does this add extra catches for errors or bias in the code, it also encourages collaborative design in AI. For example, many coding languages that are Open Source allow users to improve and edit the language.

Open Data is a movement in the field that pushes for the release and publication of datasets, or the permission to access underlying data from websites or companies. For example, recently Google and Facebook have been working to release open access databases of deepfake videos so that independent researchers, including you or me, can find ways to use AI to identify and flag this content. To learn more about Open Data and the ways it is reforming attitudes to data ownership, you can look up the Open Data Institute, run by Jeni Tennison.

5

WHY TALK TO ROBOTS:

The Potential Rewards

I believe AI has the potential to improve human quality of life significantly. Now that I've explored what can happen if AI goes unchecked, I want to press upon you that it also has many positive attributes, and this why I'm so passionate about the world of AI and making sure its benefits are realised. Amnesty International's former Secretary General Salil Shetty summed up these positives perfectly at the first AI for Good Global Summit in Geneva when he said that we could have a future where:

> The enormous power and potential of AI is harnessed for the good of humanity, promoting equality, freedom and justice. It's a future where open-source AI allows innovators across the world to harness the power of technology, where explainable AI is developed and used, allowing for AI decisions to be interrogated and challenged, and with clear legal accountability systems to ensure that the rights and responsibilities of users and developers are clear.

So let's take a look at the potential rewards of AI.

Reward 1. The world of work will be revolutionised so we can skip the boring stuff

Time, and how we use it, is something we really obsess over in our culture – and with reason. Our lives have never felt busier: there are increasing demands on our attention, we travel more and are cultivating and maintaining wider circles of people. Throughout the ages, the ultimate promise of technology was that it could be used to speed up processes and so free up our time. As a result, people could have more time to spend on creativity, innovation, problem solving, caring and even leisure. I don't know what your week looks like, but it certainly hasn't happened for me. Many of us are buried in emails, glued to our screens and drowning in notifications without really getting any more work done. The good news is that as AI proliferates, technology might finally live up to its initial promise and truly free us from the drudgery to which we've become accustomed.

Pedro Domingos writes about a vision of this in his book *The Master Algorithm*:

> In the future, you'll have your own AI, and it will hold all your data and share it only as needed. And you'll have a job that can be done neither by computers, which lack common sense, nor by unaided humans, who have only so much time and memory. AI is a horse for your mind, and horses don't compete with their riders; they let them go farther and faster.

For an office worker in the future, it is very likely that they will be paired with an AI in order to do their job. as the human being, our task will be to ensure that the combination of us and the AI makes the best employee. The AI will have all the answers because the Internet's knowledge will be at its fingertips. But our responsibility will be to ask the right questions. Kathryn Parsons gives us some great insight about this in the next chapter. Ultimately, those who have developed a way to lean on machines will accrue more time.

One reward of AI is that when it comes to work, it promises to do away with the over-complicated. Part of this will come from the abundance of voice-activated machines being used. Where once machines came with manuals and instructions, AI foresees a world where trial and error will be the way to learn and subsequently make a system work for you.

Take teachers, for example, who are often weighed down by administrative tasks that can get in the way of spending time with students. Imagine a machine that could automate the scoring of tests or streamline finicky tasks outside the classroom. This would free up time that could be spent planning and delivering lessons, ones that are better tailored to the needs of their class and the needs of individuals. A machine cannot understand the psychology of each young person in the room in the way a skilled teacher can. This is a teacher's USP.

As this example shows, as certain tasks are delegated to machines, our time will be freed up to be spent in more creative and interesting work. Jeanette Winterson talks about this in her interview in the next chapter. But we do have to be aware that there needs to be a huge amount of reskilling to thrive alongside an AI-dominant world, and we must consider how to maximise our existing skills to fit within this framework. I'll talk about the practical side of this a bit more in Chapter Six.

Reward 2. New and exciting jobs

We've seen the positives of what will happen when AI automates elements of existing jobs. The second exciting part of AI revolutionising the workplace is that it will also pave the way for the creation of many new jobs, including those we cannot yet begin to predict. The advent of AI in the workplace isn't just for the tech sector. So please, take heart! Although I can't entirely predict the wild and wonderful new jobs of the future, hopefully the list below will give you some inspiration for those that AI might unlock.

Non-technical jobs in the tech sector

You'll notice that the phrase 'you don't need to be a technologist to work in AI' will come up time and time again in this book. AI will create many other jobs that work alongside the engineers and computer scientists. The number of non-technical roles will become crucial as we try and build machines that think and act like humans. There are so many steps to building a product that aren't technical. English might have been your favourite subject; you have a tonne of knowledge about different characters, narratives and what it means to tell a believable story. You could be the perfect person to write scripts for chatbots, build characters for AI games or train an AI to write original content.

Tech jobs in the creative sector

What if you were a software engineer or data scientist, and you wanted to apply your skillset to the fashion industry? There are more opportunities to do this than it might first appear. Robots are beginning to be used in manufacturing and sewing, but fashion brands are also using AI, training them on images of past collections, to experiment with generating new designs.

Or say you're into data visualisation but you want to find a cool new way to tell stories. There's a growing field called data journalism that combines the power of data science with journalism to create ways of sharing information and breaking down news using graphics. Right now, this isn't using AI techniques, but in the future AI will be used to transform this sort of storytelling.

New new jobs!

AI will also offer up a whole host of completely new and exciting opportunities for work outside of any field with which we're currently familiar. How does Virtual Reality Travel Agent sound to you? You could help design and curate the perfect holiday for someone – using virtual reality.

With so much data being collected about what we do every day, there's bound to be an opportunity for personalised memory digital scrapbooks. We could have jobs as Memory Minders. Remember, the thing about machine learning is that it finds patterns that aren't readily available to us. Memory Minders would be able to use the data from the algorithm to connect the moments in our lives that on our own we would never have drawn together. I would love a Memory Minder for the symptoms I experience after eating certain foods, in order to see patterns and find allergies. I know some of my friends would also like one to flag when they are falling into a difficult romantic relationship – but for now we have wine and friends!

Or how about a Digital Diplomat? Your job could be to navigate the ways the Internet is managed across borders. This would be a critical job because currently the Internet works all over the world, but it's normally regulated at the level of the nation state

(GDPR is an example of this because it only exists for the EU.)
A Digital Diplomat would be tasked with negotiating the rights
and regulations across borders and figuring out the best ways to
make AI and emerging technologies work for all.

Or will you spend your days fighting the climate crisis? Perhaps
you'll be a Climate Custodian. In the future, there will be many
jobs that involve keeping down carbon emissions and cleaning
up air pollution, and we'll need as many brains as possible
focused on keeping the planet alive.

I'd quite like to train to be a London farmer and try to grow my
own cucumbers. In the future, for cities to be sustainable, as many
of us as possible will need to grow much more of our own food
and will probably do so using a combination of sophisticated
hydroponics, AI and vertical farming methods.

Human-only jobs

I believe that jobs with high human interaction will become the
most sought after, and therefore higher paid jobs. If you could
get an AI physiotherapist and an AI financial advisor, which one
would you pay extra to in order to be seen by a human? Even
Andy Haldane, the Chief Economist of the Bank of England,
agrees that there will be a future where Emotional Intelligence
(EQ) rivals IQ, and jobs like nursing, caring, education and
leisure are rewarded. Meanwhile, good news for those who are
makers and crafters: Haldane was also clear that the artisan
feel is already commanding a premium and objects that are
handmade could become an even more luxurious badge of
honour. This will of course increase price for manufacturers, but
also for software and online services where accepting you'll pay
more for a human-in-the-loop could become the norm.

Reward 3. Fighting Climate Change

Greta Thunberg, teen activist, has been making the public more aware about the dangers of climate change. According to a 2018 IPPC report, we have until 2030 to prevent the planet warming by an additional 1.5°C. This is a huge political, business, cultural and technical challenge. AI technologies can, and already are, playing a role in the fight against climate change.

Firstly, researchers can use AI to create models and projections about weather, such as what area might be prone to flooding, and the impact of that flood on the eco-system and community. This can be helpful in pushing for climate justice by targeting policies that are most likely to have the biggest impact: if researchers can predict which areas are most at risk, then they can concentrate their efforts on prevention. AI can also help measure environmental change through satellites and image recognition.

AI is also part of the toolbox for reducing carbon emissions in our day-to-day lives. Take transportation for example. Two thirds of global energy-related CO_2 emissions are generated by road users, and to combat this, AI is being used in autonomous electric vehicles, for ride-sharing apps, as well as in more complex systems for smart transportation.

AI technologies can be used to manage energy demand, so it can be predicted when there will be an energy surge. This helps balance supply and demand on the energy grid, reducing the amount of power we waste.

There are also great things AI can do when it comes to changing behaviour – food waste is a great example. Companies are beginning to use AI technologies like image classification to track food waste, especially in the hospitality industry. The hope is that by figuring out what food regularly goes to waste, companies can change their purchasing habits to be more sustainable.

The 2030 Agenda for Sustainable Development, adopted by all United Nations member states, provides a shared blueprint for peace and prosperity for people and the planet. At its heart are the seventeen Sustainable Development Goals (SDGs), which are an urgent call for action by all countries – developed and developing – in a global partnership. A recent report published in the scientific journal *Nature*, and authored by an international team of scientists, found that AI could help achieve 134 different targets from those seventeen goals. Specifically, they found AI could make huge inroads by providing quality education and fostering sustainable cities and communities. You can research these goals further, and it might be a great starting point for a project at school or to take to a company you work for.

I could keep giving more and more examples here because there are so many exciting ways AI has the potential to start to address climate change, and by the time you read this, I hope it's progressed further. You can keep in touch online with what people like Emily Shuckburgh and organisations like Open Climate Fix, Matter More and Climate Action are doing. Tech experts from across the globe are uniting to put their combined skills towards tackling this issue. If you are interested in this space, follow these organisations, and others like them, on Twitter, and find a way to reach out and get involved.

This reward can only be reaped if tech companies curb their own impact on the environment. For example, electricity is needed to power the computers that solve complex maths, and which enable AI systems to arrive at an answer. The more data, and the more complex the maths, the more electricity is needed. Emma Strubell, from the University of Massachusetts Amherst in America, has with colleagues estimated that the carbon footprint of training a single AI is as much as 284 tonnes of carbon dioxide. The transformer we used earlier to write the next verse of a Spice Girls song used the equivalent of five times the lifetime emissions of an average car. It's not as simple as this, though. In many cases, the upfront cost of building an AI solution is heavy on electricity, but afterwards this can be used many times without being so intensive. This is definitely a space to keep an eye on!

Reward 4. Helping us to live healthier lives

Remember when I wrote about the outlandish claims in news headlines about how AI can cure diseases? Well, although these claims are often inflated, AI does indeed have the potential to help us live healthier, longer and happier lives. AI can be applied to many areas in healthcare: robotic-assisted surgery, virtual doctors, improving administrative processes, simulating clinical trials and diagnostic devices. It will range from everything from risk prediction, to new discoveries of causes, signs, cures and manifestations of disease. My friend Maxine Mackintosh is an expert in this field and we'll hear more from her in the next chapter, but meanwhile let's delve a little deeper into some of these rewards.

- Detecting and diagnosing diseases: Recently, an AI was trained to interpret MRI images of hearts in four seconds. It takes a cardiologist thirteen minutes to do the same thing, so this is a huge leap forward for detecting cardiac irregularities.

It can be difficult to diagnose breast cancer from mammograms, but recently researchers have used machine learning to try to improve the accuracy of diagnosis. The hope is that this will become an important tool in early detection and diagnosis of breast cancer – hugely important, as early detection increases the survival rate.

- Drug discovery: Some companies have started using machine learning to try and design new molecules that could be helpful for new medications. Using AI speeds up the process of developing new molecules, and while it's a promising start, it will still need to go through the rigmarole of clinical trials and testing before the medications would be ready for the public.

- Improving services: In the last couple of years, hospitals around the world have started using robots as part of their administrative staff. The idea is that if doctors and nurses can outsource their administrative task to robots, they'll have more time to provide human quality care, which is proven to be a critical part of a person's holistic recovery.

- Facilitating independence: There is a new app called Pocket Vision that serves as a visual aid for people with impaired vision. It uses a mix of camera and text data to help the user understand visuals that are in front of them.

- Providing emotional support through chatbots: One of the areas where AI technologies have made great headway is chatbots. Not all, but some, chatbots use AI and NLP (natural language processing) to have a conversation with a user. A common example would be in customer service, but they

also have amazing potential for our emotional wellbeing.
One of my favourite examples comes from Kriti Sharma and
her team at AI for Good. Working alongside local activists,
AI for Good developed a chatbot called **rAInbow** – a chatbot
designed specifically to help people who are in abusive,
controlling or otherwise unhealthy relationships in South
Africa. It works by using an interface similar to texting in
order for its users to share their experiences and hear stories
from rAInbow to feel less alone or confused about behaviours.
It can also help users to identify what constitutes dangerous or
harmful behaviour in their partners and will give advice about
the next steps towards protecting themselves.

There are also a number of chatbots aimed towards youth
mental health, and these can be especially important because
stigma can be silencing, and these chatbots provide a non-
judgemental and confidential space.

Reward 5. Tailoring Education

AI has enormous potential to personalise how children are taught.
As well as freeing teachers up to spend more time with students,
AI has tools that grow or build on existing digital learning
systems. This allows for learning to happen online at a larger
scale than in a classroom. AI can also make education more
accessible and personalised. To give you an example of how AI
is changing education, we can look to the work of Priya Lakhani,
the founder of Century Tech. Century Tech's mission is to use AI
to make education personalised for every student. As lessons are
completed on their platform, their AI system identifies strengths
and weaknesses and then harnesses this knowledge to create
unique education pathways. Maybe you're quick with maths,

but grammar takes a little longer to stick. For your classmate, it's the opposite. The idea with Century Tech is that the AI would recognise this and respond differently to students in a way that best helps them to learn.

Reward 6. Managing dangerous jobs for humans

AI robots are particularly useful for handling what's described as 'dirty, dangerous and dull' work. Think about the task of shutting down a leaky nuclear reactor, cleaning sewers or defusing faulty electronics. Robots don't get offended, they are cheap to repair when they get 'hurt' and nor do they get bored, so they're perfect to be sent where few people want to tread.

Reward 7. Increased accessibility

AI has much scope for accessibility tools. Whether this is automatically determining whether a website or digital product is accessible to eye-tracking technology for people with physical impairment or translating tools that can provide live text of lectures or speeches for those who are deaf or hard of hearing. I can imagine how useful voice technologies like Siri are for those who are visually impaired to send texts, choose music and quickly access many other apps on their phone. Although this is a great start, there is still a way to go even with these speech-recognition technologies. They tend, for example, to recognise male voices best – especially those that speak 'BBC English'.

One reward that is close to me is the ways AI is being used to make life more accessible for dyslexics. As I mentioned in the foreword, I have dyslexia and know first-hand that I learn aspects of communication and language in a different way to other people, one that is not always included in the terrain of mainstream education or work. In the last few years, there have been so many exciting apps and technologies that have found ways to use AI to help people like me, including AI that can help identify dyslexia in children, help people with dyslexia be hired and make more creative teaching devices for teachers. I also rely on Grammarly and SwiftyKey to write coherent sentences!

Reward 8. No more waiting in a queue

This isn't about a better life but maybe an easier one. Those moments when you think – thank goodness for that app. Looking through the apps on my phone, I realise how often I use AI to speed up a process or simplify something. When talking to my mobile phone company, I've found using a chatbot has been a refreshing experience that reached a resolution quicker than usual. Online delivery services, which in so many ways rely on AI, can be a really positive intervention for people who live in rural communities or have mobility issues.

6

AI AND THE CORONAVIRUS CRISIS

In the final month of writing this book, the coronavirus pandemic profoundly altered the fabric of modern life. Like everyone else, I'm not sure how to wrap my mind around the enormity of the situation. I'm muddling through, trying to find a new normal, whilst transforming CogX into a virtual festival, supporting the AI community in my role as Chair of the UK Government's AI Council – and caring for a seven-month-old baby. Without any formal childcare, I'm lucky that my partner and I are able to share looking after our son, but that hasn't stopped me from feeling guilty about the work I am doing, and the work I'm not doing. I realise that I'm one of the lucky ones who can work from home, albeit with Otis climbing the walls.

During this strange time, as the media is flooded with personal reports from the frontline, as well as from the home, I've been continually reminded of all the different ways that women – and particularly women from BAME backgrounds – are more impacted by lockdowns and pandemics than men, and why both their health and social and economic status face a greater risk. I jumped into the online world, a place I am lucky to be comfortable working and socialising to connect with my friends and experts in health and social care. The chapter below was written quickly with the women I have at the end of a Zoom, and in no way is it exhaustive, but it captures a snapshot in time and provides advice that I believe will stand the test of a pandemic.

When it comes to health, women are on the frontline of the pandemic: 70 per cent of the health workforce caring for those affected by Covid-19 are women, and there are more women working in social care than in any other sector. Outside of the healthcare setting, women's health – both physical and mental – are also more at risk. During this pandemic, reported incidents of domestic violence have risen. The UK's largest domestic abuse charity, Refuge, received a 700 per cent increase in calls to their helpline in a single day.

On the economic side, there are more women employed in insecure labour than men, such as zero-hour contracts – a gulf that is continually growing. Women are also usually the primary caregivers, which makes it more difficult to work from home, even if you're able to do the job digitally. An interesting indicator of this is how, six weeks into the lockdown, editors of academic journals started noticing that women were submitting far fewer papers across a multitude of fields, from astrophysics to philosophy of science, while their male counterparts submitted up to 50 per cent more.

But we can't gloss over how these health and economic outcomes will impact different women's livelihoods in different ways. Poverty, prejudices, race and other existing inequalities which run rampant in our society make certain groups more vulnerable to the dangers of a pandemic. Just two months after Covid-19 was declared a pandemic, there was already clear data showing how poverty and race correlate to health outcomes. For example, in the poorest parts of England, the death rate for every 100,000 people was 55 per cent, compared to 25 per cent in wealthier areas. Many healthcare professionals, care workers, bus drivers and cleaners are from immigrant and ethnic minority backgrounds and are now dying at disproportionate rates. At the time of writing, just one example is how 68 per cent of NHS staff who have died in the pandemic so far came from BAME backgrounds. The Centre for Disease Control in the United States has released information showing that black populations are disproportionately affected by Covid-19. As we saw in Chapter Four, it's important to remember that statistics about race and poverty work together to tell a story about who is most impacted by pandemics. More so than individual biology, it is systematic inequalities that make people more likely to fall ill.

The Director-General of the World Health Organization (WHO), Dr Tedros Adhanom Ghebreyesus, has shared with the International Gender Champions network his view that 'The response to the Covid-19 pandemic must be gender-sensitive and responsive. Not only are women and children some of the most fragile population groups whose needs can be overlooked in health emergencies, but 70 per cent of the health workforce caring for those affected is also made up of women. WHO is committed to using a gender lens to continuously evaluate and improve our response efforts.' I was so glad to read this statement, and I've been working to make sure that this gendered response to the pandemic expands into the AI sector.

So where *does* AI feature in this pandemic? I felt it was important to explore how it might be used, and how many companies might increase their engagement with AI technology as the need to automate activities increases. As with all uses of AI, the rewards are not without risk, and the AI community is divided about its use in a pandemic, especially when it concerns gender equality and our right to privacy.

Health Impacts

The adoption of AI technologies in healthcare for this particular crisis will take some time to be implemented at scale. There are many cautionary tales about AI systems that have incredible accuracy in labs, but the results can't be replicated when they're deployed in real life. I wrote about this in Chapter Four, looking at health issues when we're *not* in a crisis; imagine how devastating an unreliable AI would be during a pandemic. This time between lab and the real world allows us a crucial opportunity to consider how gender will play a role before it's too late. Let's take a look at a few ways that AI is being used.

1. **Predicting:** BlueDot, a Canadian company, has built a tool that uses AI to continuously review more than 100 data sets, including news, airline tickets and climate data, in order to detect the virus early.

2. **Triage and Screening:** Many hospitals, faced with staff shortages and an overwhelming intake of patients, are using automated tools to quickly figure out who is sick with Covid-19 and who is not. For example, Qure.ai retooled an automated chest X-ray AI to help speed up detection of

Covid-induced pneumonia. Moreover, with the virus putting stress on other areas of the hospital, experts warn that there could be a risk of cancers going undiagnosed. As a result, companies like Kheiron Medical, who use an AI system as 'the second reader' alongside human radiologists to read breast scans, are helping to enable hospitals to reduce the number of experts involved in breast-cancer screening, and so hopefully speed up diagnosis.

Administration at care homes has traditionally been very paper-based, but by automating processes, and using machine learning tools like LifeLight – a camera that records vital signs such as heart rate and breathing rate – care workers, who might not have had much experience in this field, are given extra confidence when taking these vital measurements.

There is also the potential to start using AI to allow some patient to be treated at home. Online GP consultations rose from 12 per cent to 42 per cent in the first month of the pandemic, and NHS England has predicted that up to half of GP consultations could move to remote appointments. These sessions aren't currently using AI, but with products on the market like Skin Analytics, which allows patients to take a picture of a mole they are worried about and then flags the urgency of seeing a doctor, these kinds of tools can help lessen the load of already stretched healthcare resources.

3. **Discovering Treatments:** At the time of writing, AI companies are already announcing that they have isolated drugs that could help with treatments. DeepMind, whose COO is Lila Ibrahim, used its deep-learning system to make

predictions about the protein structures of coronavirus, which cause the disease. The system uses a machine-learning technique known as 'free modelling' to help it predict protein structures when no similar structures are available. Joanna Shields, CEO of BenevolentAI, explained to me how they used AI to identify Baricitinib, a drug already approved for the treatment of rheumatoid arthritis, as a potential treatment to prevent the virus infecting lung cells. They have entered a controlled trial with the US National Institute of Allergy and Infectious Diseases.

4. **Data Insights and Modelling:** The Covid-19 pandemic highlighted an urgent need to really understand what the data collected was telling those in charge at hospitals. A brilliant team, led by NHS directors Indra Joshi and Ming Tang, partnered up with tech giants like Cindy Rose, chief executive of Microsoft's UK division, and start-ups to address this challenge. Within weeks they built a secure datastore that helped central and regional decision makers across the country gain insights into what was actually happening on the ground, and therefore make more accurate predictions of where vital resources such as life-saving equipment would most be needed. They brought together data that has never been collated before at an unprecedented speed; they achieved results in just a few months, rather than the years it usually takes for projects like this to materialise.

Social Impacts

Lockdown has increased the use of a variety of online products and platforms, many of which use AI technologies.

1. **Online Learning:** Many schoolteachers and university lecturers have had to move their teaching to online platforms. Century Tech, whose CEO is my friend Priya Lakhani, uses AI in its online learning platform, where teachers can set and collect homework. These are auto-marked, providing students with immediate personalised feedback, and the data gathered gives teachers insight into each student. It was one of the first companies to offer its services for free during the pandemic.

2. **Delivery Services:** I imagine that you will all have a favourite delivery service, and I can almost guarantee that each journey they take is being optimised by AI. I can tell the difference between the companies that are using the data efficiently and those that aren't by the number of times I see the same delivery driver going backwards and forwards on our street, repeating journeys in a random order! I was excited to hear that Chinese autonomous vehicle start-up Pony.ai has launched a self-driving delivery service in California, as it's a giant leap for the technology, but I was sad on a personal note to think about how quickly AI self-driven delivery might proliferate, because I have really grown to appreciate and enjoy talking (at a distance) to the local delivery drivers.

3. **Communication**: With the ban of social gatherings came a mass movement of online hangouts and business meetings. Even *Vogue* switched its platform by riffing on Virginia Woolf's seminal novel and launching A Zoom of One's Own – a video hangout to discuss modern life in isolation. Many video-meeting companies use AI to give their customers a better service: Zoom offers automated note-taking for meetings, and Microsoft Teams leverages AI to filter out typing, barking and

other background noise so that people can happily tuck into lunch on a hands-free call without munching down the line!

While you were reading the above, I wonder if you spotted all the potential unintended risks associated with some of this awesome technology? It's probably clear how, in many ways, AI will be an important tool in the fight against coronavirus. But I hope you will have seen that in our haste for these AI technologies to be rolled out, discussions about gender could be skipped over, and the issues I exposed in Chapter Four could rear their head in unintended ways. This could lead to technologies not working as well for women, or – even worse – having an increased negative effect.

By now you know that, historically, AI systems are biased against women. So can we rely on them making decisions that are inclusive for all genders during a pandemic? One aspect of the Covid-19 pandemic that is becoming increasingly clear is that men who contract the virus are more likely to die than women. So what does it mean to triage symptoms or find drug candidates using AI when we know the illness itself impacts men and women differently? We need to remain vigilant when we read about these breakthroughs, and look at how training datasets were created, who was included and how did they normalise for gender?

The social impact is similarly concerning. We already know that women have traditionally had lower levels of tech literacy than men, and we've also learned about the digital divide – who can access and use the new technologies that are built with AI. Therefore, during this pandemic, we need to ask who is able to go online and actually benefit from AI-powered resources for learning?

The ability to work from home varies significantly depending on what kind of job you do. In May 2020, the UK government published a report which showed that over 50 per cent of people who work in the tech industry can work from home, but there are many other industries where that statistic is less than 10 per cent.

And what about our exit strategy? Data-driven technologies are likely to play a vital role in enabling societies to transition out of the crisis, restart their economies and return to 'normal' life. If AI technologies haven't taken gender into account when they were being built and tested during the crisis, the modelling of how to leave the lockdown won't be representative either.

On a positive note, we have already seen that if issues are flagged and raised, then we can change the outcomes. For example, there's been a lot of coverage about personal protective equipment (PPE) being designed for a tall male body and therefore not adequately protecting female healthcare workers. Caroline Criado Perez and others have used their platform to draw media attention to the problem, and hopefully this means that PPE provisions will start to change.

UN Women (United Nations Entity for Gender Equality and the Empowerment of Women) are working with the WHO to ensure Covid-19 data is disaggregated by gender and age. Grass roots organisations are also putting pressure on governments. For example, Data for Black Lives have created an action that highlights racial disparities in infection and death across the US. It also names those states that have failed to make this data available. Indra Joshi, who I mentioned earlier as the creator of the Covid chest X-ray database, told me: 'It's really important we acknowledge how diverse our population is. We're working hard to ensure the database contains diverse and representative images so the outputs are useful to the many not the few.' And like many other parents and carers during this time, she's juggling the joys of work with a five-year-old dancing in the background!

In the days to come, there will be a huge shift in social behaviour as tracking our own health becomes the new normal and an essential part of protecting women for a future labour market. We'll be encouraged to do this not in order to get fit or look sexy, but because health data will identify vulnerability to the virus, in addition to tracking whether you have had symptoms of Covid-19 or any future viruses.

But this health tracking also poses ethical problems for women. For example, we already know that women between the ages of 25 and 49 are penalised in the workplace due to bias held by employers who deem them a 'maternity flight risk'. Of course, it's illegal to discriminate on the basis of fertility or motherhood, or any aspect of a person›s health, but it's important to consider how it might be used unconsciously against women. So it's imperative that the systems we do use safely track and maintain privacy to ensure that any data collected doesn't further divide citizens into categories or classes.

I spoke to my friend Amy Thomson, who is in the thick of this debate right now. She's the CEO and founder of Moody Month, an app that supports women by tracking their hormones. Many potential investors expect her to have a revenue stream from selling this data. She makes clear that even if brands *did* want to know who is most likely to get pregnant in the next few years, she'll never sell this data to a third party that wants to advertise to her members. She told me: 'We anonymise all data today, but most importantly we think ahead to how we can better protect women globally from data vulnerabilities. Female financial independence is vital to a future of gender equality, and building technologies that can support women in achieving this, not polarise their vulnerability, is all about data privacy.' This is the sort of attitude that I would hope for from all app developers, but sadly that's not the case.

So my biggest piece of advice to you is this: keep your wits about you. Eleonora Harwich, Director of Research and Head of Tech Innovation at Reform, has passed on this important warning: 'You need to be careful when agreeing to use products, even in a very regulated market like healthcare. Creepy usages of data can be legal because an app might rely on your direct consent to something rather than a blanket safety for all users being used. Regulation can't protect you from the effects of bias before it has already impacted you. Be astute and inquisitive before installing anything.'

I hope this chapter has given you enough information to feel more confident when engaging with the tech that will be rolled out to ease the pandemic's impact on society and the economy. I haven't yet decided how I'll respond, but I'm lucky that I have the friends and the context to make an informed choice. Ultimately, you can seek out technology that helps to guarantee your privacy, or accept riskier tech for a reward, or something in the middle, just as long as you do so knowingly!

7

THE INTERVIEWS

Through my work at CogX, I've been fortunate enough to meet many women who've informed how I think about how to talk to robots, and I've collated a cross-section of them here. You'll see that some are hard-core coders and engineers, but the majority I want to introduce you to are women who, like me, came to technology through other disciplines like fine art, history, maths, philosophy and fashion. Traditionally, the skills these women hold have been described as 'soft skills', but it's these that have been at the core of their successes working and thriving in a tech and AI world.

'*Get rid of the notion that you have to be a technologist to have technological conversations.*'

ALIX DUNN

ALIX DUNN

ALIX DUNN is an entrepreneur who is hired by organisations from all over the world to help them understand the frameworks needed to build responsible and ethical technology. If something involves the intersection of technology and society, she's there.

Alix coined the term 'technical intuition', and when she first explained it to me, I felt as though the clouds had parted and I finally understood a part of myself that I'd been struggling to verbalise. She helped me to feel confident and capable of making personal and political choices about technology without feeling as if I needed to learn to code. I wanted to introduce both Alix and the concept to you because I genuinely believe that everyone has this intuition and she gives some great advice about how to unlock it and develop it.

Okay Alix, first things first – what is technical intuition?
It might help if I start by explaining similar types of intuition. When we cross the street, we know that cars moving toward us are heavy, fast and a danger to our physical safety. We know that a big haul of groceries isn't going to fit into a tiny, flimsy bag. We know that being kind is the right thing to do, even if we also know we struggle to be so sometimes. These intuitions have developed over millennia and through our lifetimes, based on direct experiences, education and social communication. The rapid embedding of digital layers to our physical life has huge

implications for how we relate to our world. We don't expect fellow humans to get a psychology degree to understand others nor expect to feel pressure to become an automobile expert in order to be able to navigate streets. Why should we expect them to become an engineer to advocate for their digital rights? Why should they have to become a data scientist to understand how their employment rights may be affected by datafication of their workdays? Why should they have to learn to configure a server to know how to set up an online business?

Technical intuition is the layer of knowledge and instinct we should be working to build so that our navigation with digital spaces and engagement with digital possibility is as mature as it is in physical spaces. Technical intuition is partly a term designed to highlight the value that a 'non-technical' person can bring to technical questions and to support the development of knowledge and learning that can grow technical intuition in people who will never and should never have to learn to code.

What types of technical intuition are there?

Technical intuition comes in as many shapes and sizes as there are people. Some of us want to learn everything about a particular aspect of technology so we can build an entire career on excelling at that one piece of the puzzle; others are interested in new breakthroughs and identifying ways to put them to practical use to solve real problems; others are interested in leading organisations to grow and adapt to technical change. If we start from the proposition that no one knows everything and everyone knows something, then understanding how to identify your own passions and learning styles makes it easier to seek out the input of people with other tendencies and to learn over time how to develop your strengths further. Like to manage collaborative teams to do creative things? You need to build a network of people who like to specialise. Like to lead more established organisations to accomplish more with technology? You need to find people who can help you interpret new possibilities and manage collaborations. We overemphasise strictly specialist knowledge when we think about who has technical capability. But technical specialists in particular need support to succeed. When what you know a lot about is technology, then the technical intuition you are working towards is the capacity to incorporate other expertise through effective collaborations. Technical intuition is knowing how to make choices about the technology we use, how to engage with that technology, knowing when we should be angry about technical systems and failed policies, and how to ask questions and pursue more understanding as technology changes.

How can we all develop technical intuition?

First, get rid of the notion that you have to be a technologist to have technological conversations. Then think about your current technical conceptual scaffolding. What do you know about technology that you can further explore, and what areas of technology might connect to it? What excites you about technology, and what support do you need to translate that excitement into real action? Over time, the goal is to build more scaffolding for new concepts and to find ways of learning the lingo that can unlock your ideas. Meet people with different types of expertise, so you can learn how to ask good questions and develop your own path of learning. Eventually, as with any second language, you can start dreaming and imagining in technology. What might you create? How might you explore the underside of a system by learning more about how it works? The key is to see any new learning or experimentation as growing your instincts, and to build your own path. You are not a passive consumer of technology – you can shape it and make it your own.

Why do you think some women feel they aren't technical? What barriers have been put up?

To me this is closely related to the idea of imposter syndrome and fetishisation of technical 'experts'. Imposter syndrome is a funny term. It implies the individual experiencing it is flawed. But for women engaging in technical fields, there is clearly something else at play. Women are discouraged in school, at work and by society. It is no wonder that this would lead women to question themselves. We continue to undervalue the skills that make or break a project, choosing to respect bravado over competency and collaboration. And we elevate a self-perpetuating (and incorrect) perception of technical fields as the purview of men. In AI, all of this is made worse by the nature of the field. For AI businesses to be successful they have to trade in pure overconfidence, and the race to build AI businesses that can accept investment has led to a rush to hire narrow experts. Because of the challenges women face at every level of a technical career, this further benefits the pool of men who are positioned to succeed in a marketplace that overvalues a specific type of expertise.

How can we change the system to be more inclusive? How can we build up women's confidence in this area?

The onus shouldn't be on individual women. The problem isn't that women aren't confident, it's that the community needs to change and become a place where women leaders feel comfortable starting new things in this area – saying I'm here and I have something to add – rather than accepting the status quo and figuring out how to break through it. Ironically, because of social marginalisation and pressure to adapt to a hostile status quo, I find that women are better at making design decisions that accommodate complex needs – and what more complexity could you need than when designing AI systems? I think that the social conditions women have been forced to navigate has helped sharpen certain capacities. That said, I think we need to be very careful making assumptions about women's capacities and strengths, as this can lead to path dependencies that further limit what careers are available to women. Women are just as capable and powerful technically, and I don't believe we should give up on fixing the systemic toxicity that fails women who are passionate about careers and innovation in technology and deeply technical objectives.

How do you talk to robots?

I find talking to robots really irritating, mostly because of the way I've been encouraged to interact with them and because of the way they represent distribution of power. AI is sold as a shortcut for companies and others with power. When a shortcut fails, the cost of that failure is paid by individuals. And maybe we have the consumer or political power to punish the company or government that subjected us to the robot, but more likely, we are stuck in a chat window screaming REPRESENTATIVE at a computer screen. You either overestimate what a robot can do and set yourself up for disappointment and frustration, or you underestimate what it can do and then you are ruthlessly monitored and targeted. I'm hopeful that we'll get more clarity about what machines can and can't do and have more control over when we are forced to interact with machines. For now, I keep my eyes open and find ways to shortcut . . . to a human.

*'I try to show people
that embracing new forms of
technology can help them scale up,
and build a great product that
people will want to use.'*

SHARMADEAN REID

SHARMADEAN REID

SHARMADEAN REID is an entrepreneur whose stated mission is to use technology to empower women, economically, socially and culturally. She's the founder of Beautystack, a platform she built to empower beauty professionals. Traditionally, beauty professionals have not needed to engage with technology, but with consumers now wanting to book everything at the touch of a button, she created an app especially for them.

Sharmadean started life as stylist. Since I've known her, she's always captured the zeitgeist for trends. She's well-known as the founder of WAH Nails, the first place in London to get what I'd call New York manicures. People wouldn't describe Sharmadean as a tech nerd – but she is! Her journey to realising that technology could work for her, rather than her working for it, is one of the core things I think is most important for us to understand. It's also why together we set up an organisation called Future Girl Corp in order to arm women with the skills that they need for work, to grow their businesses and to leverage technology to their advantage.

Sharmadean, how do you think your childhood prepared you for a world in technology in ways that you wouldn't have realised at the time?

There were two things about my childhood that played a massive factor in my relationship to tech now. The first thing was MTV! I was obsessed. It showed me how a TV network combined with great content can inspire millions. Before MTV, you didn't know what it was like to be a boyband in Britain or a rock star in LA – MTV was a window to other people's realities in a way that we now take for granted with social media. But the most defining aspect of my childhood was attending Thomas Telford School in Shropshire. The school was run as a business with a sophisticated tech infrastructure that was way ahead of its time in the mid-1990s. There were no lesson bells – every student had a card that we used to swipe into registration and classes, and all our details were stored on that magnetic strip. We all worked on laptops and had compulsory typing and business lessons. Our unique curriculum was sold to other schools. At the time, I had no idea this wasn't the norm. It was incredibly innovative, pushing the boundaries of how technology was used in the every day, and now that's what I aspire to do.

At Beautystack, you have to introduce tech to people who haven't needed to use it in their field before. What's been the biggest challenge in this process?

Without a doubt, the biggest challenge has been converting them to the idea of efficiency. Most company founders are obsessed with productivity and efficiency, but if a beautician doesn't already engage with tech, when you try to persuade them that it'll make their life more efficient – if efficiency isn't a top priority it can be difficult. What I try to show people is that embracing new forms of technology can help them scale up and build a great product that people will want to use. And at the end of the day, if they achieve this, they'll make more money!

How do your clients talk to robots?

First of all, talking to robots is not the same as talking to humans, we're a long way from that being the case. Machines just don't have enough context – they don't know what it means to be talking to girl like me from Wolverhampton, rather than talking to a girl with a different background from another part of the country.

But this can also be part of the fun. At Beautystack, we built a simple chatbot form collector for our sign-up process. I absolutely hate filling out forms online, so I wanted to use tech to create a better service where my clients didn't have to suffer in the same way.

All the feedback from users has been that this is the most delightful part of the package! There's no AI involved, but I wrote the chatbot's personality to be just like mine and people love the experience of communicating with it – it's behaviour they understand but in an unexpected setting. This is why I feel so strongly that linguistics, philosophy and the humanities are subjects that need to be included in the manufacturing process of future AI.

What's the one piece of technology that you wish existed but doesn't?

I wish that there was a platform that could read the personalities of people I meet; like an automatic Myers-Briggs score! I'd love X-ray vision about how to get the best out of them, and out of myself too. I think AI could play an amazing part in conflict resolution if we could have an instant understanding of someone's personality.

*'Find your tribe – don't try
to go on this journey alone.'*

ANNE-MARIE IMAFIDON

ANNE-MARIE IMAFIDON

ANNE-MARIE IMAFIDON is a British computing, mathematics and language marvel. A child prodigy, aged eleven, she was the youngest girl ever to pass A-level computing, and was just twenty years old when she received her Master's degree in Mathematics and Computer Science from the University of Oxford.

Anne-Marie is a renowned champion for women and girls looking to find a role in the world of STEM. Her experience and pioneering spirit led her to co-found the Stemettes in 2013, an award-winning social initiative dedicated to inspiring and promoting the next generation of women aged five to twenty-two in the STEM sectors. Through the organisation, she channels her incredible vision for a more diverse and balanced science and tech community by running programmes, panel sessions and hackathons.

Don't be put off by the fact that Anne-Marie is a former child prodigy: she's extremely funny and one of the most down-to-earth people I've ever met. We can all learn a lot from her because Anne-Marie can always be relied upon to call out bullshit when she sees it!

Anne-Marie, you've been teaching girls to code for five years, but AI signals the dawn of an era where we might not need to code anymore. What skills do you think girls need to stay ahead?

Even if you don't need to learn how to code, you still need to be digitally literate. Think of it in the same way you would with the old schools 'Reading, Riting and Rithmetic': now we've got to add Rdigital. So when it comes to future skills, digital literacy is imperative but this doesn't mean you have to run away with it. At school, when you're learning to write, you're not necessarily training to become a poet, or a journalist but it's a skill you can't live without. It's the same with AI – you don't need to be an expert in order to participate.

At Stemettes, we say to the girls that it's about having appreciation for technology and how it works, but also being comfortable with learning, and learning how you learn. Ultimately, when we look at caring, communicating, collaborating and empathising – those human qualities and actions that technology can't yet emulate – it becomes clear that it's those skills that we need to double down on to contribute to, and thrive in, a burgeoning AI world. Whether this starts with learning from other people, how to use digital platforms, or with making time to read up on AI, this is the biggest arena where girls need to feel empowered. It's not about being a genius, or having more attributes as a human being, it's simply about knowing more than you did five years ago, and in five years' time knowing more still and continually improving in this arena. All you need for this is a growth mind-set. Learning how to drive your own self-esteem and self-confidence is also important. Technology is going to change the world, so you need to know that you have the right capacity to adapt alongside it.

***The statistic is that 80 per cent of jobs will be digitalised
in the near future, which means that almost everybody
will need to understand how to work with technology. What
advice would you give to girls who in the past have found
tech boring and scary?***

I try to show girls that technology is just a tool, an enabler, in
much the same way a pen is when you're trying to get your
thoughts down on paper. It's there to help you enjoy the things
you already like just that little bit more. My advice to girls is to
find your tribe – don't try to go on this journey alone. Whether
it's online, or in person, find people that you can learn with,
and skill-up with, and share with them what it is that you find
boring or intimidating, because technology is only as good as
the people that are using it. You yourself are the best person to
understand how technology can help you overcome things in life
– whether that's being more efficient, environmentally conscious,
or bettering your mental or physical health. Perhaps you already
use YouTube for make-up tutorials, or technology that helps with
driving skills. Maybe you help your grandparents with tech. Now
think about scaling up and how this betters each scenario. Once
you start looking at problems and how to strategize solutions,
technology does help in a way that might not necessarily be
obvious if you're only using it for Instagram or vlogging.

***What's the one piece of technology you'd love to exist that
doesn't yet today?***

I need teleportation! It's annoying that we haven't yet worked this
out yet. I have to travel so much and the idea of skipping security
at airports or train stations would be amazing. Short of that, I'd
love to see technology that makes going through the whole
security process a lot easier. Hurry up guys!

How do you talk to robots?

Funnily enough, I don't like tech that you have to talk to because I don't trust the companies that build the tech. I have very few devices that I talk to, other than my iPhone and talking to my laptop through Skype. When I'm using these, I find that I don't talk to them any differently than I talk to human beings. In my daily life, I'm methodical and logical about processes and things that I need to get done, so I make tech fit into this by automating what I can, setting up spreadsheets, or using apps like Splitwise to keep track of money between me and my partner. I like to let tech work for me in little, convenient ways that increase my efficiency. I also use a bullet journal, which is analogue, but I use it to write down which system I'm going to set up and automate, or for accounting ideas. So I prefer to type rather than talk to tech. Even for things like when you call your phone company and an automated voice intones, 'Say what you want', I usually just press mute until they put me through to an old-fashioned type/touch response!

NOTES

'*Society hasn't done a very good job of celebrating women, and we don't have enough women leaders in the first place.*'

JESS WADE

JESS WADE

JESS WADE is a British physicist at Imperial College London. Her public engagement work in STEM subjects champions women in physics and tackles gender bias.

How often do you use Wikipedia as an entry point to research? Jess knows it's likely to be often, and she also knows that the number of women in the science and technology sector has been grossly underrepresented on Wikipedia, therefore skewing people's research. To address this, she orchestrated over six hundred new articles about notable female academics to be added to the site. She is also super involved in the education sector, and represented the UK on the United States Department of State-funded International Visitor Leadership Program 'Hidden No More', and served on the WISE Campaign Young Women's Board and Women's Engineering Society (WES) Council, working with teachers across the country through the Stimulating Physics Network – including keynote talks at education fairs and teacher conferences. She supports the engagement of school students through school activities and festivals, and the organisation of a series of events for girls at Imperial College London. She may also be the only woman in the world who has walked into a nail shop and asked for a constellation on her gels.

Jess, you've done loads of outreach work around STEM subjects for girls in schools. Can you tell us what you think is missing in the education sector?

Every single successful woman scientist I've met has praised their high school teachers – the physics teacher who taught them about star formation, the chemistry teacher that showed them flame tests, the maths teacher who guided them through complicated proofs. But the UK has an extraordinary science teacher shortage, so the majority of young people aren't so lucky – maybe the people reading this book didn't have a specialist teaching them, and as a result were put off certain subjects for life. The UK was the first country in the world to bring in mandatory computer science lessons for primary school kids, and we still haven't got the expertise to teach it properly. As you've learned in this book, not all of the exciting digital innovation – or AI – takes place behind a computer screen. I think the most exciting research – and education – happens at the interfaces between different subjects. But without technically excellent teachers who are able to go beyond their syllabuses, interdisciplinary work is restricted at schools. Another problem is how early we make young people specialise – we should spend our teenage years working out who we are, not selecting the three or four subjects that are going to define us for the rest of our lives. We aren't very good at career guidance, so unless you, your parents or your neighbours know someone working with tech or data, the first time you realise you might be good at it might be as you turn the pages of this book.

If you were back at school what would you do to prepare for talking to robots?

I spent my childhood making PowerPoint presentations, playing Zoombinis or The Sims and building websites. My brother and

I set up a hotel in our bedrooms, and we'd intercom each other from different parts of the house to book a room, then one of us would check an Excel spreadsheet to make sure it wasn't occupied (it was never occupied). As new tech was introduced, we'd learn it at the same time as our parents did – can you remember sending your first text message? Mind. Blown. I guess that was a pretty good introduction to not being afraid of technology.

If I were back at school today, I'd try and introduce the concepts discussed in this book outside science and technology classes. With children documenting every moment of their lives on social media platforms, we have a responsibility to talk to them about who owns that information and how to both protect themselves online and benefit from technology. Instead of encouraging children to keep their phones hidden away, I'd try and start discussions about why we're so addicted to them in the first place, as well as thinking about how we can use them to benefit learning.

Why do you think women aren't recognised in their fields?

It's a twofold challenge – one is that history and society haven't always done a very good job of celebrating women and the other is that we don't have enough women leaders in the first place. It's only in the past one hundred years that women have won the right to vote, own property and graduate from university – and certain scientists are still trying to prove that women are intellectually inferior to men. I became much more of an activist after reading Angela Saini's *Inferior: The True Power of Women and the Science that Shows It*, which shows how bias impacts the way we design and interpret scientific experiments – and how our patriarchal world has intentionally written out the voices and experiences of women.

Having said that, I think we're getting better – I genuinely believe it is the most exciting time ever to be a young woman. You've got heroes like Greta Thunberg, Artemisa Xakriabá, Alexandria Ocasio-Cortez and Simone Biles to turn to for inspiration.

How can other women help change this?

In the Obama administration women would repeat and amplify each other's ideas until they were eventually picked up – and I think that's a neat way to avoid missing women's ideas. I would hope that everyone reading this has a cheerleader of some kind – someone who'll look out for and support them (maybe it's your best friend? mum? aunt? high school French teacher?) and connect them to opportunities and celebrate their successes. But if you don't: find one, and become one for someone else. Tell your best friend about this book and encourage them to tell their friends too.

How do you talk to robots?

I'm a big fan of crosswords and frequently work *with* Siri on the tricky clues – that's not *technically* cheating, right? Actually, there are many types of labs that benefit from scientist-robot interactions; from robotic pharmacists that can quickly locate the correct prescription to automated chemical reactions and analysis of extraordinarily large data sets. Outside the lab, I try and teach robots – mainly through contributing to Wikipedia. Content on Wikipedia is indexed almost immediately by Google – you've probably noticed that Wikipedia pages float to the top of any of your searches. That means that Wikipedia pages feed Siri, Amazon Alexa and Google Home. As home assistants increase in popularity so too does our responsibility to make Wikipedia as accurate and neutral as it can be. But at the moment, almost 90 per cent of Wikipedia editors are men – mainly white men in North America. And they create content about what they're interested in – battleships, historical sites and football players. But they don't create content about women – actually, women make up only one in five biographies on the encyclopaedia. For the past couple of years, I've been working on making Wikipedia less sexist by writing biographies of women and people of colour, so that the next time you ask Siri to name a famous physicist it doesn't automatically reply with the name of a white guy.

Maybe a better question is how do I *not* talk to robots. A couple of years ago my friend was in a play called *Privacy* in New York. It taught me about the extent to which our precious data is scavenged from us – from Location Services tracking our every move to search engines selling our requests to advertising companies. I can still remember the thousands of people leaving that Broadway theatre unchecking various options in their phone settings.

When should we talk to robots?

We should talk to robots when it benefits us – maybe that's delivering humanitarian aid to a dangerous location, navigating your Uber out of a traffic jam or monitoring your flock of sheep using an app. We should talk to robots when we have got better at understanding what we want to get out of the conversation, and when we have more sophisticated ways of knowing whether it is indeed a robot that we are talking to at all.

'The most important thing is equipping yourself with the right skills and mindset to prepare for changes in technology.'

KATHRYN PARSONS

KATHRYN PARSONS

KATHRYN PARSONS graduated from Cambridge University with a degree in Classics and started her first business at twenty-six years old. She is the Co-Founder and CEO of technology education company Decoded. Decoded run Data Academies for organisations such as Nike, M&S, Unilever and UBS, rapidly up-skilling talent in emerging technologies such as machine learning. They work closely with boards and leadership teams worldwide decoding complex technologies, enabling them to lead in a digital age, and are on a mission to fill the data skills gap and realise the positive potential of technology within the economy.

Kathryn is also a regular international keynote speaker on 'The Education Revolution', sharing what she believes governments, CEOs and individuals can do to give themselves permission to learn and be part of the future economy. She campaigned successfully for Code as a mandatory subject on the UK national curriculum and received an MBE for Services to Education in 2014.

Seven years ago, I found Kathryn crouched under a table at STREAM, a conference run by Ella Kieran for media company WPP in Greece. She was handmaking T-shirts for a tournament of Ping Pong Fight Club. To someone else, it might have looked like a long and laborious task, but I thought it seemed fun so grabbed a pen and joined her under the table. The adventures have continued since then! Starting businesses, campaigning for change, we are united by our passion for making things happen, our belief that technology can be an instrument for positive change and our urgency that women need to be a driving force in establishing our future policies.

Kathryn, you studied Classics at university. Can you tell us a bit about your journey into coding?

I studied Latin and Ancient Greek at Downing College, Cambridge University, and somehow found myself ten years later working with code. From a young age I loved learning languages. I studied Latin, Ancient Greek, French, Italian, Mandarin and even invented my own secret languages. For me, languages were codes waiting to be deciphered. Languages enabled me to create new conversations and relationships but most importantly, they revealed new worlds and the cultures within them. For me, code was just another language, but one touching the lives of billions and creating the future digital world around us. I wanted to learn.

The first barrier when I started was that there was nowhere to learn. The second barrier was that everyone told me I couldn't. I was too old, I hadn't studied computer science . . . I was a girl. My response was to create a place to learn: Decoded. And most importantly I gave myself permission to learn. The rest is history. Always give yourself the permission to learn whatever it is you want to.

At Decoded you now run Data Academies, which upskills people in advanced data analytics. You describe this as 'putting powerful tools in people's hands'. Can you tell us a bit about the women who attend these courses?

Yes, the new suite of data tools available are indeed powerful. Machine learning and artificial intelligence have the power to solve problems in ways that were previously unthinkable. Decoded are putting these tools in people's hands, enabling them to supercharge the work they are doing. Women are brilliant problem solvers so I am excited we are working with such incredible professionals across a range of industries. Moreover, we need to be solving more problems in the world through the lens of women. We recommend for our clients to ensure that 50 per cent of attendees are female. It can be a challenge when we are working with male-dominated industries, however, the Decoded brand has always attracted disproportionately more women that the technology industry standard. It is a gender neutral, modern and welcoming brand, which is refreshingly different when you think about traditional 'technology education'.

What skills do you think are most useful to learn today?

Employers are looking for four things when recruiting. A learning mindset, speed, an ability to solve problems and technology skills. It doesn't matter which languages and tools you choose to learn, just be mindful that technology adapts and changes. The most important thing is that you embrace the lifelong learning journey and adapt to the changes as they occur. Traditional qualifications no longer cut the mustard for employers, they are looking for more than just exam takers and rote learners. Some are even doing away with academic credentials as an entry requirement at all.

Which jobs do you think have become more accessible in the past few years in ways that excite you?

Being the CEO. The female entrepreneurship revolution is just beginning. It presents such a huge opportunity for women and also for economies as a whole. A vast untapped well of entrepreneurial potential has yet to be fully realised. Have the confidence to take your ideas and turn them into businesses. It is okay to feel a little bit scared. Create the future you want to see through the medium of a business. Infuse it with your own culture and values and most importantly create products that bring joy into other people's lives. It's a learning journey you won't regret.

Do you think AI will change the types of jobs out there?

The categories of careers aren't going to disappear: people will still be accountants, musicians, doctors and artists. People will still have a need to care, to learn, to entertain. But these jobs will evolve, and large parts of them will be automated and technology will be integrated. The most important thing is equipping yourself with the right skills and mindset to prepare for these changes. Ask yourself, what industry do I want to get into, how do I ensure that I am at the technical forefront of it, am I learning as much as possible and harnessing new tools to solve new problems? If you are, the wind will be in your sails.

This all sounds brilliant. What's your advice for women reading this book if they want to engage with AI? What's the best way to change your mind-set?

Here are my four steps to begin:

1. **Give yourself permission to learn**
   It's never too late. It does not matter if you didn't study computer science. It is for girls. If I can do it, you can too. Don't listen to the naysayers. Give yourself permission to engage with one of the most important topics and technologies of our age.

2. **Create an environment conducive to learning**
   You need time. Four hours per week minimum. Use them to read, access online courses and meet-ups.

3. Buddy up and don't do it alone

Find other people going on a learning journey and learn together.

4. Find a problem to solve!

Technology in isolation is meaningless. Find a problem to solve and try to use technology to solve it.

How do you talk to robots?

At Decoded, we're teaching people about data science so we're engaging with robots every single day. We 'talk to robots' by translating (decoding) what they do into human language. Conversely, we mentor human beings so that they can interact with robots. Essentially we are teaching people how to problem solve using technology and we examine a vast array of problems ranging from disease diagnosis, fault detection, food distribution, waste reduction and beyond.

At Decoded we fuse the human with the robot. Data, algorithms and coding languages are somewhat useless alone. The robot needs the human. And the human needs the robot. Somehow the human has been made to feel like the least powerful part of this equation but this isn't the case at all. Humans + Robot = Problem Solving Machine of the Future. We are on the cusp of an interesting shift in human beings' ability to solve problems using technology. What we really need to be asking ourselves is what problems are we asking the robots to solve, are they big enough and are they the right problems we need to solve for the future we want for ourselves and the next generation?

'*Robomorphisation is about starting to think of humans as machines.*'

BETH SINGLER

BETH SINGLER

BETH SINGLER is a junior research fellow in Artificial Intelligence at Cambridge University, exploring the social, philosophical, ethical and religious implications of advances in AI robotics. She has a background in social anthropology of new religious movements and digital identities. As a part of her public engagement work, she has produced a series of short documentaries. The first, *Pain in the Machine*, won the 2017 AHRC Best Research Film of the Year Award. She's written numerous articles, appeared on BBC Radio 4 and at the Hay Festival and contributed to exhibitions about AI at the London Science Museum, the Cheltenham Science Festival, the Barbican, the Being Human Festival and the Cambridge Festival of Ideas.

Beth was the first person who, for me, succinctly explained how humans and machines are morphing into each other. These are concepts that have been running around for many years, but she helped me understand how humans start to adapt as a direct result of the machines with which they interact. Not only are machines changing, *we're* changing. She calls this robomorphisation – a crucial concept to wrap your head around.

***How are we humans changing ourselves to talk to robots?
Is that something we should be doing, as part of evolution,
or are there safety precautions we should put in place?***

There's more going on than just how we change our behaviour to
suit automatic systems, although that is a significant cultural shift
that we can observe with specific examples. At its most obvious
and pernicious level, this shift involves things like dealing with
algorithmic biases that can't recognise our diverse natures. It can
also involve humans dealing with even more bureaucracy, as we
have to reform our presentations of ourselves to suit simplistic
systems. This might occur when content creators find 'hacks' to
deal with the algorithm's decisions on what's relevant, or when
job applicants have to rewrite their CVs to suit an automated
decision-making system's preferences.

But there is a deeper, more dangerous, cultural shift also going
on here. The counterpart to the anthropomorphism that we are
bound up in when we talk to robots is 'robomorphisation': when
we start to think of humans as machines. There's a very long
history of reductionist and metaphorical thinking that relates
human nature to the technological perspective and innovation of
the era: the hydraulic system, the electrical circuit, the factory, the
machine, the quantum, etc. In the case of the view of the human
as machine, we can trace the idea back to the philosophical
thinking of Julien Offray de La Mettrie and René Descartes. But
in the modern era, it's more than just a philosophical stance or
metaphorical device. Robomorphism can have serious, real-
world implications: the more we interact with machines the more
we are inclined to view humans as machines and treat them as
more replaceable, more enduring and less valuable than they
actually are. The Amazon workers striking for proper breaks and
better treatment literally had banners saying, 'I am not a robot.'

However, just as it's hard to stop people's tendency towards anthropomorphism, robomorphism is a difficult trend to push back against. When discussing anthropomorphism, people comment that we're 'hard-wired' to see the non-human as human, another technological metaphor appearing even in our language about anthropomorphism. What we can do is address the social implications of this view of the human and insist on a humanistic approach to the employment of humans, and we can be very careful about the trajectory of the future of work itself as automation suits particular financial and political interests.

Why do we always see a 'Man VS Machine' rhetoric when people think about AI? Could AI and humans work better when they are paired as a team?
The 'man vs machine' rhetoric, and its louder more terrifying cousin, the 'robopocalypse' or robot uprising, both reflect a lot more about human nature than the truth about the technological sophistication of AI at this moment and what it's going to do to us. We, through anthropomorphism, are inclined to see minds in places where they are not, and then we react badly to what we infer about those minds' intentions based upon what we know about human minds. We know we want to be free – therefore any AI with a mind wants to be free. We know that we can be violent, petty, mean, etc, so obviously AI minds will be too. Throughout history we have enslaved human and non-human others, and we assume that the AI would prove to be very similar to its 'parents'. These views play out again and again in our science fictions, but also appear in the popular presentation of AI in the press.

There are of course those who are also very excited about working with AI to enhance and complement human skills, such as Garry Kasparov who was beaten at chess by DeepBlue in

1996. In 2018 he said that, 'I lost [chess] but I survived, and
I thought if you can't beat them, join them. From now on we
have no choice but to work with machines and make the best
algorithms.'

What got you excited to work in this field?

I've always been a bit of geek, and even before I started
considering the stories that we tell ourselves about AI, as a
post-doctoral researcher and anthropologist I'd spent a lot of my
early years watching and reading stories about robots. I think
the intersection of our imagined futures and the actual future
that appears to be crashing upon us with the emergence of
seemingly disruptive tech, presents us with a very interesting
time to be thinking again about some of the larger questions. I'm
not a philosopher, so I focus a little more on the question of how
we talk to robots, or how we imagine them. There are so many
possibilities for reflection on humanness, and I'm excited to know
what you think and see if we are right or not in our assumptions!
This moment is a fascinating one in which to be researching and
thinking.

How do you talk to robots?

I actually don't talk to robots that often . . . I used to have
Cortana on Skype, and my other half and I would ping her
suggested responses back to each other, but they were so flat
and monotone that I started shouting at Cortana – and she shut
herself off in protest. Or did she? More likely I just changed a
setting on Skype and didn't realise. But I think it's interesting
how we do interact with 'robots' and infer intentionality when
perhaps it isn't there. I was actually invited onto BBC Radio 4
once to discuss whether we should be polite to AI assistants.
Unfortunately, the interviewer got into asking me about when we

would have Artificial General Intelligence instead of asking me about interacting with robots. But what I would have said is this: even if the technology is nowhere near a stage where there is anything like a mind or consciousness, how we behave with other non-human beings reflects on our own nature. So being impolite or mean to robots tells us something about our own innate humanity. Of course, that reflects badly on me because I did get cross at Cortana for being useless, but I never said I was a good human!

Which book or books would you advise people interested in this to read?

Many books I read are 'primary sources' – views from within the communities I'm researching, so they provide insight into the kinds of narratives influencing our view of the robot and the human. These texts include work by Bostrom, Kurzweil and Moravec, and they give a perspective of how we imagine robots. My work would also not make sense if I didn't continue to be a geek and watch and read a lot of science fiction, and I would recommend some, likely already familiar, works in that genre such as *Black Mirror*, Asimov's short stories, *R.U.R.* by Karel Čapek and *Silently and Very Fast* by Catherynne M. Valente.

'We could use AI to enhance sex –
to bring people pleasure or help
them connect with others.'

KATE DEVLIN

KATE DEVLIN

KATE DEVLIN is Senior Lecturer in the Department of Digital Humanities at King's College, London. Like many of the brilliant women featured here, Kate didn't begin her career in the tech sector. Instead, she trained as an archaeologist before moving into computer science, and now her research is in the fields of Human Computer Interaction (HCI) and AI, investigating how people interact with and react to technology.

One of the reasons why I wanted to introduce you to her is because she's a driving force in exploring the field of intimacy and technology, moving beyond how we talk to robot, to how we form physical and emotional attachments with them. She ran the UK's first sex tech hackathon in 2016, and has written about sex robots in national press and in the phenomenal book *Turned On: Science, Sex and Robots*, which explores how the emerging and future development of sexual companion robots might affect us and the society in which we live. Whenever I'm asked about sex robots I used to blush and come out in confused-feminist hives, but then I met Kate. I'm very grateful for her serious but wickedly funny breakdown of the intersection of AI and sex.

**Kate, you've spent a long time thinking about how we
form emotional attachments with robots. What do you
think are the risks and rewards involved with this?**

We've seen dramatic – and wonderful – changes in the way we
connect with each other through technology. When it comes to
robots, it's a whole new social category and we're really only at
the beginning. Robots and AI (and in particular, conversational
AI) could be a way for us to find comfort and companionship that
might be missing from our lives. That's not to say it should *replace*
human contact – I don't think anyone wants that to happen. But it
could be a way for us to feel included and cared for, in much the
same way as we turn on the TV or radio when we're home alone,
to have the sound of other people around us.

There is concern that conversing and interacting with robots
could make us in some way deficient when we then interact with
other humans, but I think a lot of that is fear. It's the same fear we
see around automation: that we'll lose control of parts of our lives
that are important to us. In fact, we're pretty good at knowing the
boundaries between social categories. We don't really have a
basis for worrying that emotional attachments to machines will
ruin the way we interact with humans. As humans, we cope well
with technological change (whether that's the printing press or
the Internet) and we adapt to it.

**To what extent do you think intimacy can and should
blossom between humans and robots?**

We can definitely project our feelings onto anything that has
traits of being alive, whether that's assuming that animals have
complex emotions (they might!) or that a chatbot is flirting with
you (it could be, some day). I would never deny that someone's
feelings are simply imagined. Feelings of care or love don't

have to be reciprocated to be valid. So yes, I think we can feel intimacy. Should we, though? Maybe in some ways that would be beneficial: Dr Julie Carpenter's work on soldiers and their bomb-disposal robots showed that there was an emotional bond formed. That bond is very useful given that the soldiers' lives (and lives of others) depend on the proper functioning of the robot. I don't think there's any reason we shouldn't feel an emotional or intimate bond with a robot or AI – provided that bond is not harming us or making us feel bad.

In your book* Turned On *you explore the advent of sexbots. Do you worry that women will be further objectified by more naturalistic robots that exaggerate existing stereotypes?

I am concerned about the representation of people outside of the white, male default when it comes to robotics. With sex robots, the current prototypes are reductive stereotypes of the female form. That's not helpful. It's because the current design came out of the sex doll market, where the idea hinged on creating a 'perfect' sexy replica of a Barbie-doll-like woman with exaggerated proportions. I think it's also a terrible approach because we are terrible at making human-like robots. It's technically incredibly complex and expensive, and not believable. I'm not sure we'll ever see a robot that you could mistake for a real human. It's a technological dead-end.

Will AI change how we have sex?

Maybe! We have wonderful technologies that we could build into experiences and wearables that learn from what we like and give us immersive and sensuous interactions. Or perhaps we could have AI partners: RealDollX, the leaders in developing sex robot technology, have an AI girlfriend app (and are developing a boyfriend version) so you can carry your virtual girlfriend with you on your smartphone or tablet. In Japan, the Gatebox device – a glass cylinder housing a hologrammatic anime character projection – will greet you, converse with you and flirt with you. There's no AI in it yet, but the company that produces it wants to integrate it.

Overall, we could use AI and associated tech to enhance sex – to bring people pleasure, or to help them connect with others. I don't think it will fundamentally change that very human act, but it might make the experience better for some.

NOTES

'AI can do amazing things for health, but it doesn't have all the answers.'

MAXINE MACKINTOSH

MAXINE MACKINTOSH

MAXINE MACKINTOSH is a researcher in health data science, working between the Alan Turing Institute, University of Oxford and the Health Foundation to make NHS data work for everyone. She's co-founder of One HealthTech – a community that champions and supports underrepresented groups, particularly women, to be the future leaders in health innovation. The community has over 12,500 members across over fifteen 'Hubs' globally. At the time of the interview, she was a Churchill Fellow and was travelling across east coast America and east coast Africa to hear about the latest activities, initiatives, products and communities in health technology.

When I first heard her lecture about data bias, she commanded the stage, put the world to rights and absolutely blew me away. She was more stand-up comic than academic and had the audience howling with laughter. We subsequently bonded over a joint love of fashion and of sliding memes and gifs into our presentations. And we both wish for a Harry Potter-esque Room of Requirements.

I knew she was the perfect person to speak to about both my concerns and hopes around AI and healthcare.

***Maxine, as women, are there areas of AI's use in
healthcare that we should be particularly careful about?***
In some respects, healthcare is the ultimate AI sector, with
its big data sets and multitude of guidelines and patient
pathways. Healthcare is highly institutionalised, codified and
almost always gets defined at the population level. Based on
epidemiological studies, which look at the rates of conditions
and who gets affected and why, significant decisions are made
across governments, companies and hospitals to inform the right
course of action. This can range from which drugs are approved
to what's the threshold for screening. This is nothing to do with
AI really, but to do with bog-standard 101 statistics. If something
on average works for a population, for some people it will work
really well, and for others it will work less well. As is always the
issue, what happens if those people for whom it consistently
works less well are one specific group, for example, women.
When you throw AI into the mix, this gets even harder to pick up.

Here are two angles in healthcare where things can start to go
wrong:

Strict processes in healthcare help to reduce variability in care
and ensures where possible, the best decisions are being
made for patients. But what happens if those processes are
wrong, or wrong for some people? There are countless studies
looking at gender disparities, some of which are based on
misrepresentative historical data, and some of them are as a
result of codified practices of biased clinicians. For example, for
no known biomedical reason, studies have found that women
can be 50 per cent more likely to die due to preventable blood
clots compared to men. Women are less likely to be referred for
knee replacement than a man based on equivalent pain score.

And women suffering irritable bowel syndrome are more likely to have lifestyle changes suggested to them, yet men are more likely to receive X-rays. Gender discrepancies are everywhere in healthcare practice, so if we blindly automate as much as we can based on current practices, we are indefinitely entrenching these variations behind a piece of software, which is likely to remain unchallenged for a long time.

But how does the data used for research or to inform guidelines end up 'wrong' in the first place? In the case of gender differences in clinical trials, research carried out suggests it's for three reasons: Firstly, science likes to replicate previous science. If a study has been carried out before with specific findings based on a population of young white men (which up until recently was most trials), then in order to see if, say, your drug works better, you'll want to keep your population as similar as possible to the one before to see if your drug performs any better. Secondly, it can be very burdensome to engage in research. It often requires coming in and out of a trial centre, testing out a new drug, having invasive examinations, you name it. It was found that women are much less likely to engage in these studies due to caring and dependent responsibilities. Thirdly, women have been often excluded from trials due to substantial monthly fluctuations in hormonal levels. So instead of researching how these changes may affect how drugs interact with women's bodies, often a simpler solution has been to exclude women altogether.

Is there hope? Absolutely! We are moving towards 'precision medicine', which in part means moving away from thinking about populations. It means building targeted solutions for you alone, based on your unique make-up, behaviours, genetics and preferences. In order to do that we need lots of data, the ability to link the data up, and analyse it in almost real-time. So being a woman, or minority group will at some point not make a difference because if you are in a trial, then the trial is just you or people like you.

You told me that what really makes us sick isn't the actual disease itself, so how can AI manage the other factors at play?

Less than 10 per cent of your health is actually made up by healthcare. You are healthy because you have somewhere to live, friends and family who love you, you can afford to eat (and eat healthily). Throw a bit of your genetics in there and that's mostly what makes up whether you are sick or healthy. This is a famous framework called 'The Social Determinants of Health', coined by Sir Michael Marmot. In some respects, using AI for highly specialised, hospital-based problems somewhat misses the point if the ultimate goal is for AI to help us have longer, happier and healthier lives. The issue is that those highly specialised hospital-based problems are often specific and data-rich, both of which lend themselves well to using AI. But by the time you are in a hospital, receiving that crucial care, it can be too late.

Until recently, the data that sat in those 'Social Determinants of Health' was hard to come by. But as more and more of our seemingly mundane daily lives are being digitised, these data may hold the answer. For example, mobile phone data that can show how far you move from your home has been found to identify people with mental health problems. Your food shopping data may show how high in fat or high in sugar your diet is, or even more interestingly, how your palette changes, which is an early symptom of some neurological conditions. Or simply, the fact your commute is from one affluent area to another may be enough to predict most of your potential health problems. This data is held by the big and new tech companies of this day; they are the new digital bank, the Amazons and Googles, or the new watch or phone. Some of them don't know it yet, but these non-healthcare companies are the healthcare companies of the future.

So AI can do amazing things for health but it should also be clear by now that it doesn't have all the answers. All bodies are really different and often don't exhibit symptoms in the same way, and we're still a long way off working out how or why.

What advice do you have for women when it comes to using products that use AI?

Product development in medicine used to be very long (as in, fifteen years long). I'm referring to developing drugs here, which can take up to twenty years and cost over £1 billion. It takes so long because of the step-by-step clinical trials and regulatory processes. We are only just starting to understand what and how we should be regulating products with AI, but as you can imagine, regulating something that's constantly evolving isn't easy. Products are being directly released into the population making all sorts of claims about health, and new medical device regulations are only now starting to catch up. Maybe not as extreme, but it's a bit like releasing a drug into a population and doing a clinical trial 'in the wild'.

As a result, the responsibility falls even more on women to think about the apps and products in their hands and hospitals. They have not had two decades of testing, so women need to ask themselves and the organisations, what datasets were these products developed on? They need to feed back to these organisations if and when their signs and symptoms do not fit the norm. They need to check and monitor the outcomes of female patients using the services, and they need a seat at the table at the point of inception. It is regularly said that if men experienced childbirth, by now it would be risk and pain-free. Whether that's true or not, the femtech market has exploded recently with everything from AI-based breast pumps and pelvic floor exercisers, to period and menopausal symptom trackers. Personally experiencing something in health is not a prerequisite to building an AI-based solution, but it sure as hell helps to personally experience jumping on a trampoline and weeing yourself a little bit because of your weak pelvic floor to know there's a problem.

NOTES

'We don't want the future
to become a story about boys
and bots.'

JEANETTE WINTERSON

JEANETTE WINTERSON

JEANETTE WINTERSON is a fiction and non-fiction writer and a Professor of New Writing at the University of Manchester. Published in eighteen countries, Jeanette writes searingly about love, desire, gender, the body, transformation and science. She has a distinctive, lyrical style that combines elements of the fantastical with a deep commitment to honesty. Her novels include *Oranges Are Not the Only Fruit*, *The Power Book*, *Sexing the Cherry* and *Frankisstein*, and are the recipients of many awards, as is her non-fiction.

Jeanette and I met for the first time over tea at the National Portrait Gallery. *The* Jeanette Winterson, I thought, the author of *Oranges Are Not the Only Fruit*, the person who wrote the GCSE text we all fawned over. It was like something out of a movie as we put the world to rights over scones! I'm very honoured to say we are now firm friends and united in our mission to shout honestly and optimistically from the rooftops to get girls engaged with AI.

Jeanette, in your book Courage Calls to Courage Everywhere _you ask, 'Is AI the worst thing to ever happen to women?'What are your greatest concerns if women don't learn how to talk to robots?_

Over the last hundred years in the West, all the prejudices and barriers that were wrongfully in the way of women have begun to disappear. We've seen women succeed in every arena and become part of society instead of living on the margins. What worries me is that if women don't learn to talk to robots, we may accidentally exclude ourselves from the tech future. Too few women are going into engineering or computer science, which means women are neither building the platforms for this new technology nor developing the programs. There's no reason why we shouldn't get involved; we know our brains are perfectly capable. It's not like the bad old times when the medical condition 'Anorexia Scholastica' was invented for women – an eating disorder that was thought to occur from too much mental stimulation! We know that women can do maths, and they can code, and they're good at engineering. What's more, you don't even need brilliant maths to do engineering – boys with B-grades go into it and women who are as capable, if not more so, don't. I've been reading about why women do or don't decide to go into tech, and I'm troubled by the men who write that women will only ever make up 20 per cent of the workforce because 'we just don't want to do this stuff'. Whenever people start saying 'women don't want to do this stuff', there's usually two possible answers. One is that there _are_ real gender differences and women are never going to get on board, which seems to me to be madness because we've already proved that in the past so-called gender differences used to exclude women were found to be more about male prejudice than female capacity. If we decide not to put the blame or responsibility on women (it's your brain/hormones/

babies etc), then we have to ask if, at every level, a culture still exists that is putting off girls and women from going into AI – whether it's the frat-dorm nightmare of the coding room or tech start-up, or earlier, at school, where girls feel they're being shut out because boys take over the subject, and teachers themselves often pander to gender stereotypes. That said, there are a lot of women out there trying to tackle this problem head on, at school and at work. Social enterprise **Code Like A Girl** is a good example, plus there are good initiatives at Microsoft and Amazon, and I'm hoping that there are enough women out there who can influence girls in time to make up the numbers in tech, AI and machine learning.

In fact, women don't have to be coders or work in machine learning. There are plenty of jobs opening up in ethics and in communication. Simply becoming more fluent in tech is important; think of it as just a new language to learn. Also, you don't have to be that good at coding if you do decide to learn it! Lots of men do it, and aren't ace. They're average. We get better by *doing*. Girls need to remind themselves that they can do this!

Is this concern partly what encouraged you to develop your writing about people and sexuality to include machines, as you do in your latest book **Frankissstein?**
The genesis of *Frankissstein* came from a book I wrote in 2000 called *The Power Book*, which explored what it would be like having virtual identities on the web, and how that multi-self would be enormously freeing. If you're a writer, you always have multiple selves, you're allowed to have the voices in your head and people pay you for it – it's fantastic! So in the early days, I thought virtual identities would be liberating for people because it would get us away from binaries e.g. I'm a boy, you're a girl,

and fixed identities, and then everyone would be able to start using their imaginations more. Most people narrow themselves into smaller and smaller spaces as they get older, and to me the web is a wonderful way to say, well, look, the space need not be small. Of course, this has turned into a really rather baleful encounter because now we don't always know who anybody is online, and there's this cyber nightmare of web-stalking and false identities – not the same as the playful, changeful imagined self I anticipated. There's trolling, hate-speech and people thinking that they can do anything they like, responsibility free. I didn't see that coming!

So with *Frankisstein*, I'd been re-reading Mary Shelley on the 200th anniversary of its publication, and I thought how we were the first generation to read that book rightly – it was like a message in a bottle sent to us so that we could see the pitfalls and perils of creating non-biological life forms, as well as its possible glory. Mary Shelley was only eighteen years old when she wrote *Frankenstein*: she was so prescient. It made me want to work with that story and bring it into the now to see what it would look like. Will we do what we always do when we achieve the dream? Completely screw it up?! Or will we realise that this is genuinely a new and democratic way forward for everyone? Which it could be!

What is it that you find exciting when it comes to engaging with machines? Could there be benefits for women in particular?
I'm excited about both embodied and non-embodied intelligence. By the embodied form I mean robots, and they could be great fun. If you can fall in love with your teddy bear you'll be able to form a relationship with a bot. Every kid in the

world is brought up on cartoons where a human being makes friends with a talking animal or a non-human life form, so we're already used to that idea. I see no problem there. It'll be life affirming, and great for kids having a little bot running around with them. I don't think it will lead to a lessening of the human spirit – quite the opposite.

I'm an optimist at heart, and I think that AI presents new possibilities where women could really benefit. The idea that we can use embodied and non-embodied intelligence to speed up our work life to get more done faster, just as we've always used machines, can really work for women. It could take the burden off our daily lives. If bots come along and help with childcare and all the stuff at home that women still tend to do more than men, then that's brilliant. We keep inventing things that make our lives more difficult, and I can't understand that. We have so many labour-saving devices and yet we do so much labour!

Everyone talks about the world of work as being threatened by robotics and automation. But if we examine the culture of long hours – something years of workplace reform modified – right back to the nineteenth-century Factory Acts, we see that long hours kicked back in during the 1980's a as a kind of macho, neo-liberal thing. And we have a gig economy where people need three jobs to make enough money. If we were sane about work, we would accept that a lot of time in factories and offices is wasted. Ask yourself how long is your productive day? Not your workday, your productive day. If you are a doctor or a teacher, for instance, you will have longer days, though both doctors and teachers could really benefit from bot helpers – and expect to have shorter days.

Of course, sometimes all of us have to stay late, you might have a deadline, your work might be seasonal – long hours all summer, shorter in winter – this isn't a rule I am making. What I am saying is that robotics and AI will upend the world of work, and that could be to everyone's advantage – if of course, we share the money around, and I am a big believer in universal basic income.

Women in particular could benefit. There's still concern over whether women can have a home and a family and a career – but I think it's such old-fashioned thinking. We don't have to beat ourselves up if we're not putting in ten-hour days on the job and then eight-hour days at home. We need to be able to provide the love, time, space and imagination that kids need, and relationships need. To me, love is like gardening – you have to do a bit every day. You can't binge garden or binge love. You can't say, 'I'll love you this weekend but then I won't see you for ages.' It doesn't work with raising kids and doesn't work in relationships.

It's not ludicrous or blue-sky to say that we could all end up working less but *not* for less pay. Robotics should allow that to happen. It's not magical thinking that it should reduce the kind of laborious work that people have to do to just to get by and put food on the table. We're very close now to being able to have a fairer and more equitable world if we can use the technology we're developing for everybody. What I'd hate is if this experiment with AI was to go horribly wrong, and people blamed it on the AI rather than blaming it on us. AI is not a dystopia by default; it could be brilliant.

How can someone's love of reading and writing fiction be useful for joining the AI revolution?

I'd like women to be really aware of AI. If you're not aware, then you can't be a player, and things just happen to you. Reading is a great way of staying informed – and not only informed about facts – though that matters a lot. We also need to develop our minds to be able think both imaginatively and critically – reading real books – fiction, non-fiction, poetry, develops the mind and sharpens our own use of language.

So women interested in storytelling can start thinking about how they can use stories to make other women more aware of AI. It's a simple idea but the world always needs better stories. We've seen the rise of women's writing in the last 100 years and now there are just as many, if not more, women publishing work than men. The barriers put up there were all artificial. So we've started to tell stories with different voices, and different interests, and therefore different consequences and outcomes. We don't want the future to become a story about Boys and Bots.

Also, there's a sort of madness at the moment where the discourse is 'Machines will take over the music business', or 'Machines will write the literature'. But as long as there are human beings, we'll need to hear stories from other human beings. And those stories will be warnings, celebrations, inspirations and changes of heart. It's really important that people can change in order to embrace the future. Einstein said, 'We cannot solve our problems with the same kind of thinking that created the problems.' As soon as your mind changes, the problem is on its way to being solved.

'Understanding how machines understand humans is an incredibly important step in making sure that we create a respectful partnership.'

HANNAH FRY

HANNAH FRY

HANNAH FRY is an Associate Professor in the Mathematics of Cities at UCL. Her research applies to a wide range of social problems and questions, from shopping and transport to urban crime, riots and terrorism. Alongside her academic position, Hannah has contributed to and made BBC television documentaries, appears on BBC Radio 4 and hosts brilliant podcasts, which you'll find listed in the further reading section. Online, her YouTube videos have clocked up millions of views, including her popular TED talk, 'The Mathematics of Love'. Hannah has also authored a number of books. Her latest, *Hello World: How to be Human in the Age of the Machine* was shortlisted for both the prestigious Baillie Gifford Prize for Non-Fiction and the Royal Society Book Prize.

What's the chance of us finding love? When should you settle down? How can you avoid divorce? These are just a few questions that Hannah uses maths to answer in her first book *The Mathematics of Love*. When I read this book, with Hannah's ability to make the subject not just fun but relatable, it made me fall in love with maths. Since reading and admiring from afar, Hannah has become a friend and I like nothing more than cheering her on as she challenges the male, pale and stale in the academic world.

Hannah, you've written a lot about AI-driven decision-making. Can you tell us about the moment you realised that our lives could be at risk from this?

AI-driven decision-making is not necessarily going to be negative. I like to think that we're currently at the teenage stage of this technology – by that I mean AI *and* the explosion of data collection and our increased ability to look for the patterns within it. We have no idea how far reaching it will be, so it's going to take a little bit of time to get through the spotty phase and work out how best to implement this new technology. There is, of course, a darker potential future where AI becomes a monster that we have to rein in, but it's not an inevitability!

I first realised the potential impact of AI-driven decision-making when satnav was installed on our phones in 2008. Before that moment, satnav was available to us via Tom-Tom or similar products as one-way inbound communication: the satellite simply told you where you were. But when that technology moved across to our phones, it became two-way communication. I remember being at a seminar at the time and someone said that if we reached the stage where an information system told us where everyone was, then there wouldn't be traffic jams anymore because a central system could reroute people to minimise traffic in all areas. It hasn't been very long since that moment, but we do now have apps like Waze which do exactly that – but because there are too many cars on the road, we haven't eliminated traffic! So that was the point where it really hit me that there's a digital twin, or a digital mirror image of the world that is revolving around numbers.

In your latest book Hello World you describe a future where the 'arrogant dictatorial algorithms are a thing of the past.' Can you share some advice about what we can do to ensure that this is the case?

What most people find difficult about the use of algorithms is grappling with how much is shrouded in uncertainty. We're used to living in a deterministic world. You drop a cup on the floor: it smashes. You trap your finger in a door: it hurts. The inevitable problem with AI is that it's never perfect. Sometimes it's only 60 per cent accurate. But even when it's 99 per cent accurate, it's still never deterministic. It's not the case that the AI says you have a tumour, therefore you have a tumour. We're not in that realm yet. It's more like, the AI says you have a tumour and this alerts us to some probability that you have a tumour. So completely understanding and accepting that uncertainty, and then *looking* for the uncertainty, is the one thing that can equip us to navigate this new era. We can investigate far-reaching claims. For example, I heard about software for sale that supposedly allowed you to analyse the waveforms of a person's speech and facial expressions to show up particular words that person was more stressed about saying. The idea was that you could use this on politicians to pinpoint their weakest topics or policies. If you live in a deterministic world, you might think that's a great invention. But if you're completely aware that this technology resides in a world of uncertainty and acknowledge that maybe it *can* do what it claims, but also ask: How *well* can it do it? When did it work? When did it not work? How often does it work? And what happens when it doesn't work? Well, then that's the best tool you can have against dictatorial algorithms.

Another example to which we could apply this line of questioning is when the American store Target 'proved' that they could work out a teenage girl was pregnant from the family shopping before her own father knew. That story was told in a very deterministic way. Target analysed the data accrued by their shopping via their Guest ID number, which ties buying habits to the credit card, email address or name of the shopper. The store then matched this data with historical data collected from women who had signed up to Target's baby registry in the past. They sent the girl coupons for maternity and baby clothing and cribs. It turns out that this girl was indeed pregnant, and so Target 'knew' before her father. There's a causal link within that story. But people live their lives forward and they're only understood backwards. The questions we should be asking in this scenario are: How many people were sent coupons that weren't pregnant? How many people were pregnant but weren't sent coupons? How often did that algorithm work? How often did it get it wrong – and if it did, what were the consequences? If you move this line of enquiry into more serious arenas like policing, court rooms, housing or when deciding state benefits, and you're already armed with this understanding that everything is wrapped in uncertainty, it's then that you can start to address the problems of algorithms more clearly.

In researching* Hello World *you interviewed what you called 'snake-oil salesmen' and developed 'The Magic Test'. Can you tell us more about this test and how the reader can use it?
So firstly, hands up, this is not my test. I heard it described by someone who funds AI research projects and receives absolutely loads of applications which claim things like 'We will use AI to cure cancer' or 'We will use AI to solve climate change'.

We're at this stage with AI where there are so many high-reaching algorithms out there without a system in place to ensure that what people say their algorithms can do is really true. As a result, I think we've opened the door to people making false claims and never being held to account, never having to publish their statistics to show how accurate they are. But at the heart of that problem, wrapped up in all of this, is the fact that AI is such a black box. You do something, get the data ready, and then the AI does the magic and you get the output at the end. So that step in the middle – where AI does the magic – is so obscure to most people that we imagine that AI can do almost anything.

So going back to the magic test, and this guy who was reading so many grant applications from people making outlandish claims about what the AI was capable of doing: he had a rule, and it was if you can take out all of the technical words in the document, such as machine learning, neural networks and artificial intelligence, and replace them with the word magic, if the sentences still made grammatical sense then he wasn't going to fund it. You can use this, too. Interrogate a claim by replacing some of the technical words with 'magic', and if it makes sense you should think that maybe something in that statement is fishy . . .

How do you talk to robots?

At home I have an Alexa and Echo, and the reason I have both is because they were given to me! One is in the kitchen and the other is in the living room and I would never have them in the bedrooms. I don't shy away from using technology, or from talking to robots, but I do limit the power of my machines. I use DuckDuckGo as my browser – a search engine that emphasises privacy protection and avoids the filter bubble of personalised search results, and I have ad blockers. I have lots of email addresses that I use for different products. During periods when I really need to concentrate on work, I have a series of apps on my computer that limit how much I can use it. Like Self Control, which shuts down the Internet. So I set a timer, and even if I restart my computer I physically can't get onto any social media. I also use Slow State, which is a black screen that you can write on, but you have no access to the rest of your computer. I also switch my phone to black and white and that makes a huge difference to how distracting it is. So I limit the power of my machines because I'm very aware of my own human weaknesses and wanting all the entertainment all of the time!

What I'd like to do is flip that question around, because sometimes we're limited as individuals in how we talk to machines, but there are certainly things that can be done so that technology can talk to *us* better and about how we collectively as a species talk to robots. Currently, there is an ongoing, expansive area of AI research that endeavours to make sure that when we speak to machines and order them to perform something, we're not giving them incorrect directions by mistake. A good example of this was when an AI was given the objective to land a simulated airplane with minimal faults. During the training period, the AI discovered that instead of landing the plane gently and delicately, if it slammed the plane into the ground deliberately then the force that would be registered would be so high that it overwhelmed the memory of the game and thus registered a perfect score of zero. So the AI repeatedly slammed the plane into the ground, killing all simulated passengers on board and destroying the plane! Therefore, you want to make sure that your AI and your robots are doing what you want them to do, not just what you *told* them you wanted them to do. AI and robots don't see the world the way that we do: they don't understand context or nuance. Understanding how machines understand humans is an incredibly important step in making sure that we create a respectful partnership.

'People massively underestimate the power that they have over their smartphones.'

MARTHA LANE FOX

MARTHA LANE FOX

MARTHA LANE FOX is founder and executive chair of Doteveryone, an independent think tank and charity that champions responsible technology for a fairer future and looks at how AI can serve people better. She has sat on public service digital projects and the board of Channel 4, and currently sits on the boards of Twitter, the Donmar Warehouse and Chanel.

Amongst her many extraordinary ventures, she's perhaps most famous for launching Lastminute.com. Watching a historian lead the way with a website that was responsible for a billion holidays was one of the defining moments in my tech education because it made me believe that I, as an arts graduate, could make use of the web too.

Martha wrote the playbook for my generation of women in tech. Without knowing it, she's been a huge influence on my life, with the pink streak in her hair and fierce ability to get the crux of any situation and lay it bare, I have long given myself advice based on asking myself 'What would Martha do?' She is the external face of tech entrepreneurship in the UK yet doesn't stop at that – she's also the internal supporter of other entrepreneurs, championing female leaders and shining a light where others forget.

Martha, you were in the first wave of humans learning to
talk to robots when you were setting up Lastminute.com.
How did you make the machines work for you?

This was twenty-two years ago, and the digital world was wildly
different. There was this idea that the Internet was everywhere,
and it was pernicious and that you had to bend it to your will. It
was a brave new world, and we were at the vanguard of inventing
e-commerce – just convincing people to type their credit card
details into a website was a really big challenge. Not only was I
not a technologist, I also couldn't code – meaning I had to learn a
lot about tech just to make the concept of Lastminute.com work.
To do this, I had to be super clear about what we were trying to
achieve and this meant fundamentally thinking like the user. We
were relentless, in fact, in approaching how the site would work
from the perspective of the users. We'd keep asking: is this what
we want for our customers? It was a slow process.

At the time, people were very worried about their relationship
with technology and that was the key thing for us to build on.
We didn't have the language we have now for tech. We were a
tiny team where everyone was doing everything, and there was
an oblique benefit to this because the _only_ thing you could do
was ask yourself: would my friends use this to book a holiday?
So we were on the website all the time, testing it and trying to
make it work for people like us. Sometimes we'd draw pictures,
sometimes we'd look at other websites as an example – Amazon
was the main one back then. I'd sit with the person who was
building the site and go through the flat pages one by one that
we were turning into html. It was so basic! If anyone was building
a website now, they'd look at how we did it and wouldn't believe it!

What have you seen as being the biggest change in our relationship with technology?

It was only in 2007 that the iPhone was invented. Before then, there were no smart phones; we were all just bashing away on our Blackberries. Things changed completely when people had a device in their hands that could do in one step things that had previously taken three. That to me was a fundamental shift in the possibility of everything – it's easy to underestimate the impact of it as everyone has become used to smart phones. Services have got a lot better, and all major industries have moved into having online platforms. These things sound so obvious now, but they really weren't back then.

Do you have any advice for people for staying safe now that they do have the power of technology that comes with smart phones?

I think that people massively underestimate the power that they have over their smartphones. I feel this very strongly. For example, I never ever take my phone into the bedroom. I just don't! I don't have any notifications switched on. I even turn off my emails. You can control your settings. Phones are built around engagement, like most social services around the world, but you can still control some of its elements. Turn down your screen's brightness; turn off all notifications; put your privacy and data settings on the highest level. Put your phone down and do not take it into places where you do not want your phone!

You studied classics; if you had to pick one course, would you suggest someone followed in your footsteps or chose to study coding instead?

I have mixed feelings about this. If learning to code is a route into having more curiosity, and less intimidation about tech, then brilliant. Do I think everyone needs to learn to code? No. Clearly we do need people who can, but it's going to change soon as machines are going to get better at coding than people. If we're looking at life skills, it's brilliant to have that technical understanding, and even a spark of interest or skill in that area means you'll never be out of work, so we should encourage as many people as possible. But it's not the only thing. You shouldn't feel frightened of tech if you don't have those skills. It's about curiosity, asking questions, and we really need multidisciplinary people going into the sector. People who understand history, philosophy, anthropology. The digital world is not optional – it's a fundamental backdrop to everything now so just make sure you're curious about it.

If you feel you do want to get more deeply technological then double down on it! Retraining to do something technical is never a bad idea. If you have even a slight predilection, then go for it. No matter what field you're going into, say you want to become a nurse for example, go away and learn the latest digital health facts so that you're always going to be at an advantage. That's not the only way to get involved. Rather than being subject-led, it's about being able to cope with ambiguity, fast-paced ideas and change. It's about being intellectually curious and having an alive mind. It could be that you get really deep into a subject, and I salute people who become experts. It's awesome! At the same time, my experience of sitting on boards at places like Twitter is that we need people who are also able to understand that technical innovation has always created these moments of skill and change and not to get frightened by that. We need people who can see things from a hundred-year perspective, not just a ten-year one.

I don't want to be prescriptive because all of these things need to come together if we are to build the most resilient society possible. To me it's about continuing to challenge your brain with reading and learning, and even if you're an exceptional coder or brilliant mathematician, to realise that there's a real value to having some understanding of the humanities. And if you're an amazing history student – fantastic, but appreciate that there are some technical skills and science-based parts of the world that are vital for our understanding of what's going to happen over the next decade. It's about trying to bring disciplines together rather than saying they should be separated.

If you had a one-minute ad slot during prime-time TV what would you use it say?

The only thing I'd be saying is that you may not think you have any power, but you have a lot. Think about everything you are doing and whether you could deploy it more effectively to make sure that all our resources and skills are being used to solve the right problems for the modern age. Are you working on things that are really going to make a difference to the next decade? Are you doing it with a diverse set of people in the room to make sure that you are inclusive and represent everybody? I think what's going to make a difference to the next decade is climate change, which is why I champion advances in AI to be used towards addressing this issue.

Martha's further reading: fiction and non-fiction

I read fiction obsessively. I'm a massive fan of Ali Smith, particularly her quartet of books *Spring*, *Summer*, *Autumn* and *Winter*. Her novel *How to Be Both* is one of my favourite novels of all time. I don't think you can ever beat reading a great Russian classic, and I go back to *War and Peace* time and again because it represents all of humanity.

'Part of the power lies in regulation, but a large proportion of power actually lies with the consumer.'

SANDRA WACHTER

SANDRA WACHTER

SANDRA WACHTER is an Associate Professor and Senior Research fellow at the University of Oxford. She's one of the few people working on the intersections of tech law and the ethical implications of AI, Big Data and robotics. She acts as a link between the makers of technology and the judges and policymakers who create a legal framework for it – a rare position for academics to be in.

When I first realised the extent to which bias and AI were going to impact women, Cognition X hosted an event called 'Why Women in AI?' I trawled the Internet and at the time it was hard to find women who talk about this subject in the UK. And that's when I met Sandra. I went to visit her in Oxford on a gorgeous summer afternoon, and she generously corrected some of my misconceptions and pointed me back in the right direction. She is absolutely brilliant at explaining how seemingly opaque rules and regulations are actually more straightforward than we might think, and how knowing them unlocks the door to some of our most important rights.

Sandra, what power does the current GDPR regulation give our reader that they might not know or understand how to leverage?
The main thing that the new data protection framework does is allow people to understand what data is being collected about them and how it's being used. It's a very good transparency tool.

When we interact with Facebook, Google, or Amazon, we might not be aware that we're leaving a data trail behind us. This data is collected and evaluated and stored somewhere. GDPR allows you to know about your own data because companies now have to tell you what they're doing with it, why, and who they share it with. They also have to ask you for your consent, or at least inform you about data collection – that's probably one of the most important things that GDPR can do for every European resident. It means I can go to Google or Facebook and ask them to give me a copy of that data immediately. Right of access allows me to have a very good overview of what data is collected about me, and other provisions allow me to rectify or delete it.

That's great progress but does this regulation have limitations?

Yes. Because companies spend a lot of time and resources collecting and evaluating your data, they often consider it to be *their* source and therefore their right to protect the business is more important than my right of access and privacy.

What can we do to arm ourselves against this gap in the regulation?

Regulation is never perfect; it's always a process. A law is always an answer to a problem, and when the problem changes, so too must the law. It's not surprising that it's not perfect yet, and it probably never will be perfect. Going forward, what I would like to see is more responsibility on the companies rather than the user. Currently, companies have to ask you for consent. This appears as a cookie banner, and you're asked if you want to share your data. But the truth is that not everybody reads the terms and conditions. Not a lot of people understand them, and few people have the time to read them thoroughly – you might have twenty or so apps on your phone alone, all with lengthy T&Cs. So companies are abiding legally, but the consensus is that it's maybe not the best way to gauge if people are actually happy with how their data is being used. I think we should shift the weight of responsibility of data usage onto the tech companies to demonstrate to the outside world that they have good intentions, are ethical and are handling the data in a reasonable way that is protective of privacy.

So what would give those big tech companies impetus to handle our data in a more responsible way? Is there something we can do?

Companies are starting to understand that ethics is a competitive advantage. There are interesting studies that show users start to vote with their feet if they feel like their privacy rights have been violated. These studies showed that people were very upset by the idea that if, for example, they were browsing to rent a horror movie one evening, the following week their browser was full of advertisements for other horror movies. It's an intrusion that creeps people out, and very often they will shy away from using that platform again.

There are also studies that show that if companies take data ethics seriously, and are very open and transparent, then people are more likely to engage with, or buy, their products. So part of the power lies in regulation, something that is very necessary, but a large proportion of power actually lies with the consumer. Start protesting by voting with your feet, and then the market will change as well.

Automated bias is a big risk for our readers. How can they spot this happening to them and what can they do about it?
This is one of the hardest questions that nobody really has an answer to yet. It's complex. One of the most important things is to understand what bias means and where it comes from because bias is nothing more than a mirror of society. We collect data about our world *from* our world, and so it reflects the realities of our world. Machine learning works under the assumption that when we look at the past, we see certain patterns about how things happened, and then those patterns are used to predict the future. Machine learning only works if the future looks like the past – and that's not always a good thing! Therefore, knowing that there is no such thing as neutral data or unbiased data is very important.

Figuring out how we deal with that on a policy and technical level is something that we need to address as a collective because it's not just a legal or ethical problem. Because there are so many disciplines that are connected to automated bias, we need to put our heads together to be effective at reaching our goal. It's especially challenging for young women because they have been, and still are, disadvantaged by history in various ways. We don't want certain stereotypes to be brought into the future. One of the most important things to be aware of, and to call forth in our approach, is to be inclusive and diverse. People who build products are often white men, and so they build products for people who are similar to them. This isn't necessarily a criticism: you only know the person that you are; you only grew up with your family, your friends, your own experience. You've never been another person. On some level we all have unconscious bias. But if we're trying to create systems that are directly going to affect the *whole* of society, for example using algorithms which decide which people do, or do not, get loans, or do, or do not, get hired – life-changing decisions, then you have to make sure that the systems that we're using are not repeating the same mistakes of the past and that technology serves the whole society and is inclusive.

You've spoken about breaking the black box and giving users explanations about why an AI has made a decision about something they might involve them directly. Can you talk about what this means for users?

Explanations are a very good first step towards real accountability. It's not the silver bullet that solves everything, but it is a good step to foster trust and increase accountability because then people know *why* they didn't get the job, or why they weren't approved for a loan, or even why they went to prison. Google has already implemented some of my work on 'counterfactual explanations' in Tensor Flow that I have done with my colleagues. 'Counterfactual explanations' now give users explanations about how AI has informed decisions. Now you can actually go and query an algorithm and ask them why a decision was made in a certain way. Breaking a black box like this increases accountability. If you can find out that you weren't approved for a loan because you are a woman, then you can use that information to go and kick up a storm because gender should not have played a role in that decision-making process. So counterfactual explanations can be used to combat bias on an individual level, but they can also tell you what kind of criteria were actually used in making that decision – if it was your postcode, or your income, for example. And that's something which has advantages for everyone. So even though very complex systems are used, the reasons for decisions are now also easy to understand for someone who is not a computer scientist.

It sounds like we're making great progress, but before all AI becomes explainable, what can we do to protect ourselves?

Be excited but sceptical at the same time. There are amazing opportunities out there, healthcare for example, where AI could benefit society. In other areas, AI makes things faster, more accurate, cost efficient – which is great. But we have to remain sceptical about the technology and not blindly trust algorithms. Query them, don't blindly rely on the data that was fed into it. Is it actually working as intended? Could there be bias written into it? How many tests has it been through? What kinds of precautions were taken to make sure that known problems have been ironed out? I usually compare it to how I have both an online diary but keep a paper diary too. That's a sensible approach, because although obviously an online diary is faster and more convenient, if something goes wrong then it's gone. The paper is my backup. In the same way we have to be excited but cautious about the possibilities of AI. It's important to ensure a safe failing by figuring out that if something does go wrong, it does so with the least amount of harm.

8

ACTIONS

I hope this book is filling up with biro scribbles and you have
started to see how you could implement new ideas when you get
back to school or work. The actions in this chapter will help you
navigate this new terrain.

Firstly, you've been curious enough to pick up this book. I
encourage you to stay this way and to nurture this instinct. It's
both your greatest weapon and your best defence against the
onslaught of fast-paced technological change. At the core of each
piece of advice is the notion that you must continue to question
everything. I know that by the time this book comes out there will
already be new things to embrace and to arm yourself against,
so the only advice that will truly stand the test of time is to foster
your curiosity and share the feeling of discovery. Don't keep
what you now know close to your chest – explaining the things
you have learned will not only benefit your friends and family, the
pure act of sharing will help you cement your own understanding.
No one should feel like a lone warrior. If women are to succeed
in an AI world, we need to do so together.

1. Embrace change

As a human, you are the most important part in all of this
change. You need to constantly remind yourself why you are
unique, special and more valuable than any machine – however
smart it is.

As you know by now, I think that because technological change
is inevitable, ignoring it simply isn't an option. The robots aren't
going away! So we need to embrace the change. But how? What
does it mean to embrace change when the change might take
over some of the jobs that we are used to doing ourselves? Can
we embrace change in a way that doesn't leave us vulnerable
and draws on our (often hidden) strengths?

Women have a lot to teach us about how to live and work
alongside machines. Throughout history a set of qualities
traditionally associated with women – compassion, care, empathy,
nurturing – have been dismissed or sidelined by the market.
Today, care work is either amongst the lowest paid jobs, or done
for free (mainly by women) in the home. But these qualities,
which have always been vital, are about to become ever more
necessary and much harder to undermine. Many aspects of all
jobs are going to be assigned to machines, but they can't do
everything that humans can do. Imagine you're a doctor: a robot
may be able to hold all of *Gray's Anatomy*, the medical textbook,
and *Grey's Anatomy*, the TV series I love, in it's system, predict
and detect diseases invisible to the human eye. But the one thing
it can't do is connect on a human level and offer genuine care
– something we know is key to patients' comfort and recovery.
Human empathy is something machines can't do.

Women have also developed another skill that will become vital in the coming years: staying on our toes. For centuries women have faced all kinds of discrimination and prejudice. Women have had to know how to be vigilant and resilient, to anticipate change and to read subtle cues and analyse the world for risks. They have had to stay one step ahead of 'the man'. Now, women can teach us how to stay one step ahead of 'the machine'.

What does this mean, practically speaking? Whatever field you might be working in, or have worked in in the past, or would like to after you leave school, run a thought experiment where you consider which tasks could be automated. It's important to note they might be automated using AI or simpler process automation.

Ask yourself – what could be done better by a smart machine? If you can predict where the machines have an edge, you can focus on honing the skills where the AI has no chance of competing. Why go head to head with a super computer?! Use your guiles to outmanoeuvre it instead. You can follow these steps to see where the moves might be.

1. Break down a job into a series of tasks for which you are responsible and write them in a long list. I'd use a spreadsheet, but you can use a paper if you prefer.

2. Draw a line or use columns on either side of the sheet and on the left write 'full automation' while on the right, 'humans only'.

3. Note which tasks are repetitive, don't require teamwork or creativity and add them to the left. On the right add the tasks which require you to deal with complex ever-changing unpredictable situations with many different people involved.

Example:

Full automation	Could be automated	Long time to automate	Humans only

4. Then look at the fuzzy middle. What are the tasks that might in the future be automated? Which of these do you think should be automated, and which do you think should remain solely in the human realm?

5. Consider what's needed to reach this level of automation and how long it might take. For example, ask yourself: is there enough data in machine-readable format (rather than lots of pieces of paper) to train a machine today? Be careful, as humans famously overestimate the change that will occur in the next few years and underestimate the change that will occur in the next ten to twenty. For example, thirty years ago they were expecting flying cars by now!

6. Now you can look at your lists and start to see which skills you might want to focus more on because they are uniquely defensible. You can also see the skills you might not need to practise because there are either tools now which can help you manage the tasks, or you can easily predict that there will be one day. This exercise could also be a good catalyst for the entrepreneur in you. Start thinking about companies you

might want to start or projects inside your organisations you
could kick off that could address these specific gaps.

I hope this exercise gives you a feel for how a job might change.
It is of course almost impossible to predict exactly what will
change and which skills you'll need to prepare to adapt to
that specific eventuality. What I'm really demonstrating and
suggesting is that you prepare yourself for *any eventuality*, while
all the time learning how to learn.

When I was at school, I found that my dyslexia meant I learned
in a very different way to my schoolfriends. I couldn't learn by
rote; I needed things to be visual and I could only understand
new concepts if I deconstructed the idea, and then put it back
together again in my own way. Struggling with the learning
format at school meant I would get really frustrated and was
often sent to detention for causing a nuisance. But it was also
the cause of me honing my greatest skill: 'learning to learn'.
Learning how you learn, and learning how to learn, makes life
a lot easier when you're confronting something new. Be kind to
yourself, and give yourself the time that you need to work out
how to understand new things. One of my friends only learned by
teaching something to someone else, while another had to have
all her lessons on audiotape repeat. Now we are adults in work,
it's still the same, and we are using those skills to keep learning.
Whether it's podcasts or courses, events or documentaries that
get your learning juices going, embrace them. It's not too late to
figure out how *you* best understand something and then double
down on that.

It's most likely that you'll have many roles, jobs and even careers
in your lifetime. If you are fortunate enough to make a choice,

seek work for what you will learn as well as what you will earn
– whatever job it is that you might be doing, or if, as is the case
for many women, you are doing multiple jobs simultaneously
or are a full-time mother or carer. Indeed, being able to juggle
lots of things at once, where situations are evolving and people
are unpredictable, is another thing that humans do better than
machines. Without even realising it, you might have already
developed the most sought-after skills in this new age of machine
learning!

In summary: think about where you can make changes.
Recognise that you may have hidden but valuable talents. Most of
all, if you can, let yourself learn. You are capable of a lot more than
you might assume. All of this will give you a unique advantage
over AI.

2. Talk to a robot

My granny used to say, 'Understand the question and the answer
will take care of itself.' This pearl of wisdom I now know was
adapted from Einstein. We can ask a version of this question
when it comes to AI: how do you get the machine to understand
your question, so that it gives you the answer you need? This is
something that is going to take practice. It's also something that will
require a little more determination from some people than from
others (as we've seen, machines reflect the biases of their creators
and this has an effect – sometimes literally – on who they hear).
But it is vital that we persevere if we don't want to be left behind.

At the time of publication, 200 million smart speakers have been
sold across the world and 11 per cent of UK households have

one, which I imagine means you have come across one or even have one in your home already. I don't want to promote any brands here, or unnecessary purchases, but these are a good way to practise constructing sentences and asking questions that AI machines can understand. You don't need to rush out to the shops – there is AI you can talk to in products you might already have. If you're an Apple user, talk to Siri, or Cortana if you use Microsoft and Google has an assistant too. Set your alarm to be voice-activated, or use a voice assistant to add appointments to your calendar, or search the Internet for you.

My friends tell me that they've given up on their home system, or that they can't bear that their car is trying to talk to them. My response is always to tell them: this technology isn't going anywhere. So instead of avoiding it, I try to help them make it work better *for them*. Below are some commonly asked questions:

My machine only recognises the men in my house, what can I do?

I have exactly the same problem at home! To start, all AI assistants have a 'wake word': this is the word that the machines are trained to respond to and will prompt the system to turn on. For example, saying 'Alexa.' After this, you'll need to start speaking clearly and explain your instruction. But many women have said they're super frustrated because the smart assistant only recognises the male voices in their house. You are not alone: a YouGov study showed two-thirds of female owners said their device fails to respond to a voice command at least 'sometimes', compared to 54 per cent of men. This is because the machines have been trained using predominantly male voices, which tend to be lower in pitch, meaning the voice assistants will struggle to understand higher-pitched female voices.

You can do something about this. In the settings of your Alexa, for example, you can tap 'Recognised Voices' and then tap 'Learn my Voice'. You'll be asked to speak so that the machine can start to map your pitch and tone.

It's not only about the way you speak. The YouGov research also found that women are noticeably nicer to their smart speakers than men, with 45 per cent saying 'please' and 'thank you', compared to only 30 per cent of male owners. I am in that 45 per cent: I told my partner that I'd installed a new app that meant Alexa would only respond if we added the words please and thank you to our questions. Technically, this wasn't true or possible, but I figured a little white lie would nudge things along in the right direction. But when Alexa was presented with words that were superfluous to the instruction, the request was not actioned. Having discussed this with some experts, it's clear that if you add words that aren't deemed necessary to the command then the machine will get confused. And so we returned to staccato demands. It was one of the first times I realised that how our machines are programmed will affect how we interact with them, and also how we potentially speak to each other and start learning how to talk to machines.

I'm in two minds about whether we should say 'please' and 'thank you' to our voice assistants. On the one hand, I was thrilled to hear about the Alexa child-lock feature that wouldn't wake the machine unless it heard one of these polite words. But it also opens another can of worms when thinking about humanising the machine too much. It's up to you to decide how the AI is treated in your house and now you can.

My kids think Alexa is their best friend, what should I do?

Rather than he, she or they, avoid gendering and humanising the machine by calling it 'it'. This is especially important for what we pass on to the younger generation because we need to establish the difference between humans and machines, and the language we use to refer to this technology can help us differentiate our roles.

I hate bossing a female voice around, so I stopped using it. Is there a solution?

When we were kids, my parents had the coolest satnav and we were able to choose the voice. Of course, my brother and I chose Stephen Fry's dulcet tones to guide us around. Sadly, the current market for AI assistants doesn't have as much choice, but applications like Siri do enable you to pick a man or a woman. Have a rummage through your settings and you might find some alternatives.

The thing has become a glorified speaker for my music and is starting to gather dust, can I throw it out?

No! You must keep exercising your muscle memory. It will feel odd at first but set yourself more tasks to do using the voice assistant rather than manually. I love to cook with my voice assistant. I don't just use it to put a timer on, but I ask it for conversions from pints to litres, and for substitutes to ingredients I don't have.

Try a day where you don't type your searches into your preferred search engine. I'm always interrupting the flow of a conversation to have a quick fact check of a friend's comment, or my own! Rather than typing into a search bar, try instead to use your phone's microphone to speak your query and share it with the group. It's amazing how at first this might feel frustrating and counter-intuitive, but after a while it becomes second nature.

Also check out the Alexa Skills Library in the app that comes with it for more exciting ways to test the AI. For weeks I used the Harry Potter sorting hat until finally I was in Gryffindor!

Is the machine always listening? I swear it just ordered Marmite when I hate it. What can I do?

The first method of silencing most voice assistant devices is via the manual, physical button. Some of us don't read the manual and are unaware that there are manual overrides so explore where yours is.

Secondly, you can turn off background listening by toggling settings in the apps that control the speakers. In time, these companies will need to have these choices turned off to allow you to make the choice to opt in. But until they do, I suggest you decide your own settings.

My bank and phone provider have changed to AI assistants, and I can't get what I need answered. What should I do?

I hate to say it, but you must keep trying. It's as much about using your muscle memory and training yourself as training the machine. Imagine if your job depended on you speaking to an AI system? Treat this as a primer. I now actually find it easier sometimes to ask questions of a bot than to navigate my bank's website. You'll see from the interview with Alix she completely disagrees, but I found I was able to be more direct and get to the answer quicker typing than calling.

However, the best AI I've seen in banking isn't actually an AI that interacts with you, the customer, directly but is cleverly working in the background to connect you with the right human assistant best placed to handle your request. This is the sort of use of AI I'd like to encourage companies to deploy more.

Now, it's not your responsibility to make sure the companies build the right tech, but you can have a part to play. You can write to your bank, give feedback privately or publicly and put pressure on them to improve their service in ways you see fit. See below for more on this.

I was horrified when my friend started calling Alexa a slut and she replied, 'Well, thanks for the feedback.' What's going on here, and how can I stop it?

You are right to be horrified. UNESCO reported in 2018 that female voice assistants have been programmed to 'greet verbal abuse with catch-me-if-you-can flirtation'. The only thing we can do here for now, other than reprimanding your friend, is complain directly to the manufacturers. I've seen the big tech giants listen to the press who've picked up on this and start to take the issue more seriously. Alexa won't now reply this way.

3. Protect yourself

Cars, the phones in our pockets and the electronics in our homes all have a compelling ability to simplify our lives and make them more fun. This is no bad thing, but it's important to be conscious that there is a price to pay for this convenience. As you now know, data is used to train machine-learning algorithms and has become the most valuable commodity for businesses today.

Many new companies are valued on how much data they have, rather than how much revenue is generated.

It's clear that we as individuals are not responsible for the risky development and use of AI technologies, and so we as individuals can't solve them – it comes from our data economy and structures of power. One person opting out of cookies on their Internet browser can't do anything to stop the larger issues of targeting, data breaches and surveillance.

We have a right to know what companies use our data for and this knowledge should be within our reach. Look at how much our attitude to food has changed over the last twenty years. We've become much more savvy consumers of what we eat by demanding information about what goes into our food, how and where ingredients are grown and what percentage of our daily allowance is represented in any one product. We've begun learning more and trying to understand what goes into our bodies. In the UK, each food item has a traffic-light system to help us make more conscious decisions. But currently, tech products don't come with any similar kind of warning.

To help address this, Mozilla has a website you can check out called Privacy Not Included, which shows consumers how secure products are. They've developed a set of minimum-security standards that tests each product under the guise of the following criteria. I think it's a great website but also just a good checklist when considering what products or services to use.

- Can it spy on me?
- Does it share information with third parties for unexpected reasons?

- Does it encourage me to change the default password?
- Does it have automatic security updates?
- Does it delete the data it stores about me?
- Are there parental controls?
- How has the company managed security vulnerabilities in the past?
- What could happen if something goes wrong?

I know there perhaps isn't enough competition when it comes to social networking – most people across the globe use Facebook or Instagram, which are owned by the same company. Keeping on top of who owns what, and which sites are safe, can be daunting and time-consuming. Thankfully, journalists like Sky's Rowland Manthorpe and the *Financial Times*'s Madhumita Murgia have done a bunch of research already. If you're unsure about a company's ethics, it's good practice to conduct a quick online search to see what headlines they've been making lately, before you use their services.

The challenge you face on these platforms is that there is a war on for your attention. Newspapers, TV shows, politicians and advertisers are all clambering to get your attention. To do this, they are often using some kind of AI technique to show you content that they already know, through the cookies you leave behind, you will click on, read, share or buy. When we use the Internet, each of us is accessing a version of it that has been specifically curated for us. But there are ways that you can challenge this reality. One way is to break out of the trap laid by companies who are working to keep products you already know and buy in your path. If you're conscious of this, you can upset these algorithms by actively seeking alternative sources of news and following people and brands you wouldn't usually. This way, you won't end up consuming what your feed wants to feed you. This is especially important during election periods when campaigners will be spending vast amounts of money to reach you. Going forward, consider varying your news sources so that you don't get stuck only reading and watching the same rhetoric from the same people and falling into the trap of self-fulfilling prophecies. The next thing to do is interrogate your source. Look at the person who shared this article or video with you. Is it a friend? Is it an anonymous avatar? Check the date it was posted, and then check how many followers they have. Does any of this raise your suspicions about whether the source is a real person or whether the news is real or AI generated? Knowledge is power.

If you have the ability to buy products from companies that reflect your values and are able to encourage others to do the same, then that's something extremely simple that can be really impactful. When you can, don't shy away from calling out companies that reinforce, or exacerbate, existing bias, unfairness or sexism. Being vocal about this, and then voting with your wallet, is a good way to influence multinationals and reinforce any legal regulations that they have to follow. This is where community-thinking is so important: tech giants will not make changes unless they can see it having a direct effect on their revenue.

A good way to protect yourself against these challenges is to know your data rights and how to exercise them. We looked at this in Chapter Four. As companies build more and more advanced AI, there will be updates to GDPR and other new regulations. My advice is, don't ignore or resent it – these regulations will give you new rights so ensure you use them. It's not just for your employer or company: this is for you.

4. Be part of the conversation

Dorothy Vaughan at NASA is my hero, and you heard her story in Chapter Two. Rather than feeling threatened by the machines, Dorothy embraced them, and with the confidence and foresight to predict the next wave of technology, she taught herself how to use the computers and ensure that she had the skills to stay ahead. But she didn't stop there: she then trained her team and ensured they were primed for change. Rather than being made redundant, they evolved to new jobs. And she did all this as a black woman during racial segregation.

Current statistics show that 80 per cent of large companies are looking at how to use AI to increase productivity. This means that whether directly, or because of these organisations, your world is changing. My advice here isn't necessarily going to be easy, but my challenge to you is to aim to be involved in some way with how this change happens. Otherwise there is a high chance it happens without your experiences being considered.

How this affects you is going to depend on what matters to you most, and what resources, power and influence you already have. If you're in a position where you can put your hand up and get in the room, make sure you do. At the moment, decisions about AI are being made in rooms mainly filled with white men. This is not only a problem for diversity of the workforce, it's about diversity of experience. The point of being in the room is not just about having your voice heard; it's about changing the terms of the discussion and bringing your experience and that of your peers and community with you.

I am incredibly privileged to have worked my way into rooms that I'm not technically qualified to be in. It's not always comfortable, and I've encountered a lot of well-placed confusion at my presence, but I've pushed through the embarrassment. In part, this is because I have a nose for a story, and an obsession with knowing what's going on behind closed doors, but mostly because I want to be able to communicate this to other people.

Some tips for getting more involved:

Bring your expertise: Keep your ears out for developments happening in your organisation and ask your boss if you can get involved. AI design teams will need designers, artists, journalists and people who can translate the real-world experience. These organisations will really benefit from your point of view.

Hone your tech skills: This book might have inspired you to be involved in the actual development. Check out the courses in the next chapter for suggestions of how to become a data scientist or machine-learning expert.

Set up your own group: There are so many special interest groups that get set up in organisations and schools, so why not set up one for AI? Even if you just meet in the pub or borrow a schoolroom for a few hours, try to get together with those around you that might be interested in, or affected by AI, and chat with them so you can agree a united response.

One way to do this is with **Future Girl Corp**, a group for entrepreneurs that I helped set up with my friends. If you head to worldwide.futuregirlcorp.com you'll find there are eleven sessions of content and advice on how to start your own local business community. With each lesson, such as 'Hiring', you can apply an AI lens and ask 'How can AI be applied here?' or 'What would we need to change in order to make this relevant in an AI world?' This will help you have conversations in a context that can be applied to a start-up.

Join a union: Historically, unions have been at the forefront of the fight for labour rights in the face of technological change and they will continue to play an important role as AI expands into workplaces. If you're a member of a union, you can ask about what they're doing to anticipate these changes (many have already hired specialists for just this reason). Many of you in industries most directly affected by new technology don't yet have unions. This is changing. If you're a gig worker you can check out the IWGB and take inspiration from the Deliveroo riders who use Whatsapp to organise, learn, communicate and protest against poor treatment.

5. Ensure others are heard

In this book, I've focused on the western working woman and I've been asking you to consider your individuality in relation to changing technologies. But I also want to stress that we need to think about how this will affect other people, other areas of the globe and life outside of our immediate communities and ways of living. I'm writing this book from my position as a privileged woman living and working in London. I recognise that this means there is a lot that I don't see, and it's up to me – and to people in similar positions to me – to educate ourselves.

We can't, for example, just ask, 'Is this technology for good?' because what's good for some people won't be for others. I ask myself: what are my blind spots? How can I work to address them? Whose voices have I not heard from and why? For me, this piece of advice draws from all of the previous ones and frames my attitude to how I approach AI in my day-to-day life. It's the importance of continuing to learn, questioning everything and building relationships with machines and people.

Your voice is much louder and even more relevant when it's representing more people. Once you understand how your group of friends feels about AI, try and find a way to explain this to someone who can influence some change, like your headmistress or your boss. You can use social networking sites like Twitter or YouTube to broadcast your thoughts, or you could write emails to influential groups. I know in my role as the Chair of the UK Government's **AI Council** I'd love to hear what you and your peers think.

However you proceed, there are ways for you to hold leadership accountable when you find things are upsetting, unfair, unjust or downright illegal. Check out **#TechWontBuildThis** (a movement in big tech companies fighting against unethical projects); **AI Now**, which is a good place to track the social implications of AI; **Glitch UK**, which protects against online harm; **Doteveryone**, a think tank focusing on responsible tech policy.

I decided that the best way I could support women whose voices weren't being included in the conversation was donating the proceeds from this book to **Rosa**, a charitable fund set up to finance initiatives that benefit women and girls in the UK.

Rebecca Gill, the Executive Director of Rosa, and I had a chat about their work. 'Rosa's vision is of a society in which women and girls are safe, healthy and have equal opportunities at work, at home and in public life. In 2019, we asked women across the UK "What does 'work' look like today and what could it or should it look like in the next decade?" There was a clear understanding about how technology underpins almost all paid employment today, but the role of automation was less clear cut; and how it might impact our lives in the next ten years was perhaps inevitably met with a mixture of apprehension and excitement. In many respects that is understandable, because it's anyone's guess. What is clear to us, however, is that automation will impact girls and women differently to how it impacts men, depending on what stage of life they have reached and what types of paid or unpaid work they might do. For this reason, Rosa believes women need to be around the table, part of the conversation, shaping the future. Women must be conducting the research, providing the expertise, setting out the evidence and the arguments about the ways that AI and robotics could work for everyone: for women and for men, for people from black and minority communities, for the low paid and the unpaid as well as the highly paid and privileged. Rosa wants to see women's organisations alongside wider civil society, think tanks, academics, civil servants, social enterprises, trade unions, tech experts and those funding the big thinking in this field to face this challenge from a different angle, to incorporate more voices.'

We discussed her biggest concern, that without this concerted effort, we will see the same outcomes as we always do. Girls and women, particularly those living on or below the poverty line, losing out in a very big way. That's why I'm committed to helping to increase the investment Rosa has to fund women's initiatives with the goal of making sure that this doesn't happen.

As our video call came to an end, Rebecca explained our (now joint) mission perfectly: 'As the hackneyed saying goes, if you're not round the table, you're on the menu. The impact of robotics and AI is too important to be left to a few well-paid men to make a meal of our future.'

9

WHAT NEXT?

Recommended Reading & Viewing

I hope this book has fed your curiosity and ignited the spark to read more in-depth about a particular subject. This list provides suggestions of what to sink your teeth into next; it's not exhaustive, but a cross section of tools I've used to learn, and I'll be updating it on the *How to Talk to Robots* website. Whether it's a novel to get your heart racing, non-fiction to study or a course to help you up-skill, now you can take your next step on the journey.

Non-fiction

21 Lessons for the 21st Century by **Yuval Noah Harari:** This set of essays is a great (albeit a rightfully scary) introduction to all of the urgent challenges humanity is facing today. Dealing with the likes of climate change, nuclear weapons and fake news, *21 Lessons* tries to make sense of what it means to live in a moment of mind-boggling change.

The Age of Surveillance Capitalism: The Fight for a Human Future at the New Frontier of Power by **Professor Shoshana Zuboff:** A look at how our current tech ecosystem has created an age where a small few hold unprecedented power to control the many. In addition to talking through the impact of this kind of capitalism, she also shows ways we can protect ourselves and fight against it.

AI Narratives: A History of Imaginative Thinking about Intelligent Machines by **Stephen Cave, Kanta Dihal, Sarah Dillon:** Fascinating journey from ancient Greece to the present day these authors examine the history of imaginative thinking about AI and help the reader use this past to have contemporary debates about these powerful new technologies.

AI: Its Nature and Future by **Margaret Boden:** This book brilliantly describes how AI has impacted, and continues to impact, disciplines from biology to education to linguistics. She also thinks through the possibilities of machine creativity, and what AI might mean for art in the twenty-first century.

AI Superpowers: China, Silicon Valley, and the New World Order by **Kai-Fu Lee:** In this book, Kai-Fu Lee looks at how China and the USA are building, supporting and managing AI innovation, and makes the argument that China is best positioned to become the leader of AI.

Algorithms of Oppression: How Search Engines Reinforce Racism by **Safiya Noble:** A vital exploration of how one particular set of algorithms that so many of us use every day reinforces racism: Google search.

Artificial Knowing: Gender and the Thinking Machine by **Alison Adam:** A must-read for those interested in the more philosophical side of gender and AI, Alison Adam thinks through topics like identity, knowledge, rationality and language to find and explore gaps in mainstream models of AI.

Artificial Unintelligence: How Computers Misunderstand the World by **Meredith Broussard:** This book focuses on something we tend not to think about when we hear about AI – that often, the algorithms guess wrong. What does it mean that so much of the technology we rely on doesn't work? And can it (or should it) be fixed?

Automating Inequality: How High-Tech Tools Profile, Police and Punish the Poor by **Virginia Eubanks:** This book examines how algorithms perpetuate inequality, focusing on how new technologies impact cycles of poverty in the United States. Using case studies from California, Indiana and Pennsylvania, this is a fascinating and troubling deep dive into one of the many risks of AI.

Broad Band: The Untold Story of the Women who Made the Internet by **Claire L. Evans:** This is the full story of the women involved in building the Internet as we know it today – something I touched on in Chapter Two. It's a surprising, funny and powerful history that fights against the narrative of the tech bros.

Data Feminism by **Catherine D'Ignazio and Lauren F. Klein:** Written by two brilliant professors, this book focuses on the relationship between feminism and data science. Through topics like power and labour, they explain how data selection, visualisation and analysis can all have gendered impacts.

The Digital Ape: How to Live (in Peace) with Smart Machines by **Nigel Shadbolt and Roger Hampson:** This book explores ways humans will have to balance control of machines with their own intelligence. If you want to learn more about the rewards of AI, this is a deeper dive into the bright possibilities of our augmented future.

Diversify: How to Challenge Inequality and Why We Should by **June Sarpong:** This unapologetic, nuanced book traces the different angles through which we can learn the benefits of diversity. In addition to building this argument, Sarpong examines where social structures have limited diversity, and the consequences of these exclusions.

The Economic Singularity: Artificial Intelligence and the Death of Capitalism by **Calum Chace:** This book bravely argues that within a few decades, most humans will not be able to work for money.

The Entrepreneurial State: Debunking Public vs. Private Sector Myths by **Mariana Mazzucato:** This book flips the relationship between innovation, industry and state on its head. Professor and economist Mariana Mazzucato argues with eloquence and clarity that risk-taking is, and has always been, a critical part of the domain of the government.

The Future of the Professions: How Technology Will Transform the Work of Human Experts by **Richard Susskind and Daniel Susskind:** This book predicts the decline of today's professions, looking at lawyers, accountants and consultants, and discussing the people and systems that might replace them.

Future Politics: Living Together in a World Transformed by Tech by **Jamie Susskind** This book challenges us to rethink what it means to be free or equal, what it means to have power or property, what it means for a political system to be just or democratic, and proposes ways in which we can regain control.

Hello World: How to be Human in the Age of the Machine by **Hannah Fry:** This book looks at how the way the current 'Age of the Algorithm' will change our lives, and how it already has in ways we might not even recognise. Written in elegant, relatable and easy-to-understand prose, this is the perfect guide to navigating our modern world.

Human + Machine: Reimagining Work in the Age of AI by **Paul R. Daugherty and H. James Wilson:** This book focuses primarily on how company structures are changing in the face of AI. It's the perfect read if you're interested in learning more about how different types of business adapt and adopt these emergent technologies.

Human Compatible: Artificial Intelligence and the Problem of Control by **Stuart Russell:** Can we coexist happily with the intelligent machines that humans will create? 'Yes,' answers *Human Compatible*, 'but first . . .' Through a brilliant reimagining of the foundations of AI, Russell takes you on a journey from the very beginning, explaining the questions raised by an AI-driven society and beautifully making the case for how to ensure machines remain beneficial to humans. A totally readable and crucially important guide to the future from one of the world's leading experts.

Inferior: How Science Got Women Wrong – and the New Research That's Rewriting the Story by **Angela Saini:** *Inferior* takes us through ideas in biology, anthropology and psychology to show how encoded sexism exists in their analyses. She not only demonstrates the impact of this, but also gives us hope for the future by explaining how new research is working to correct and dismantle these biases.

Invisible Women: Exposing Data Bias in a World Designed for Men by **Caroline Criado Perez:** I've quoted Perez throughout *How to Talk to Robots* because *Invisible Women* examines all the ways gender impacts data-based decisions and designs in our world. Looking at topics from health diagnostics to automotive engineering to bathroom design, Criado Perez asks over and over, where are the women in the data? And what are the consequences of our invisibility?

Lean Out by **Dawn Foster:** Maybe you've heard about *Lean In*, the book by Sheryl Sandberg about becoming a successful female entrepreneur. Dawn Foster has different advice in this book, arguing that meaningful feminist change comes not from a couple of powerful women rising to the top of the corporate ladder, but by breaking the ladder down completely.

Let It Go: The Memoirs of Dame Stephanie Shirley: In this poignant, honest memoir, Dame Stephanie Shirley takes you through her journey from refugee to business woman to philanthropist. Reading this memoir shaped so many of my ideas on the importance of creativity, growth and dedication!

Machine Learning and Human Intelligence: The Future of Education for the 21st Century by **Rosemary Luckin:** What is the future of education in the era of big data and AI? Here, Rosemary Luckin gives a thorough, easy-to-follow introduction to the ways AI technologies might shape classrooms of the future.

The Master Algorithm: How the Quest for the Ultimate Learning Machine Will Remake Our World by **Pedro Domingos:** This book pulls together digestible examples of how AI is impacting our society, ultimately thinking about what this means for the ongoing search for a single algorithm that can learn everything and anything from data.

Once Upon an Algorithm: How Stories Explain Computing by **Martin Erwig:** In his creative take on explaining how algorithms work, Martin Erwig takes readers through problem-solving techniques of characters in familiar stories – think Harry Potter and Grimm's Fairy Tales – to explain the underlying ideas behind all the mathematics.

Prediction Machines: The Simple Economics of Artificial Intelligence by **Ajay Agrawal, Joshua Gans and Avi Goldfarb:** For those interested in the way AI is changing business, this book uses the premise of 'cheap prediction' to make sense of both the uncertainty and productivity of AI. It's practical and accessible, and a good place to start if you already have some background in economics.

Programmed Inequality: How Britain Discarded Women Technologists and Lost Its Edge in Computing by **Mar Hicks:** This fascinating book tells the story of women involved in the origins of computing in the UK, and shows how their discrimination and subsequent erasure from the field impacted, and continues to impact, gender and economic inequalities in the country.

Race After Technology: Abolitionist Tools for the New Jim Code by **Ruha Benjamin:** Presenting the concept of the 'New Jim Code', Ruha shows how a range of discriminatory designs encode inequity by explicitly amplifying racial hierarchies. She provides conceptual tools for decoding tech promises with sociologically informed scepticism.

Rebooting AI: Building Artificial Intelligence We Can Trust by **Gary Marcus and Ernest Davis:** From two of the leaders in the field of AI, this book describes the current landscape of research and how far we really are from AGI.

Road to Conscious Machines: The Story of AI by **Michael Wooldridge:** If you're looking for a comprehensive history of AI, this book by Michael Wooldridge traces the development of the field in a way that is accessible, clear and thoughtful. It's especially helpful if you pair it with reading *Broad Band* or *Programmed Inequality*.

Saving Bletchley Park: How #socialmedia Saved the Home of the WWII Code Breakers by **Dr Sue Black and Stevyn Colgan:** Intrigued by the story of Alan Turing and Joan Clarke in Chapter Two? Well this book by Dr Sue Black tells a different story of the location of that fateful Second World War codebreaking, Bletchley Park. It was set to be demolished in the early 2000s, but social media saved the site!

Smoke & Mirrors by **Gemma Milne:** As we have learnt, highfalutin headlines about science and technology are nothing new. Yet the hype has a dark side. It can deceive and distract. Gemma Milne reveals how hype in AI, quantum computing, brain implants and cancer drugs, among others, can be responsible for derailing crucial progress. Hype can be combated if we can see where, how and why it is being used. This book is your guide to achieving exactly that.

Superintelligence: Paths, Dangers, Strategies by **Nick Bostrom:** For those captivated (and maybe overwhelmed) by the idea of machines with human-level knowledge, this book from philosopher Nick Bolstrom discusses what it would mean to reach that point of superintelligence.

Technology Ethics by **Stephanie Hare:** This book addresses one of the most vexing problems facing humans today: biometrics, big data, data protection and children's data. Stephanie challenges and inspires those creating, investing in, regulating and reporting on technology – and it will empower all readers to understand technology better and hold it to account.

Think Like a Programmer: An Introduction to Creative Problem Solving by **V. Anton Spraul:** If you're looking for a more technical read on algorithms, this guide is a creative way to get some familiarity with what programming looks like, and how it's written to solve problems.

Turned On: Science, Sex and Robots by **Kate Devlin:** You met Kate in Chapter Six. In her book, Kate delves deep into how AI could change sex and explores what the creation and development of sex robots may mean for our understanding of relationships, violence, attraction and gender.

***Weapons of Math Destruction: How Big Data Increases
Inequality and Threatens Democracy*** by **Cathy O'Neil:**
I've referred to *Weapons of Math Destruction* a lot in this book
because O'Neil uses clear and recognisable real-world examples
from a range of sources (like the financial crash of 2008 or student
loan applications) to show how even the best intended algorithms
can have consequences for social and economic inequality.

What Works: Gender Equality by Design by **Iris Bohnet:**
Given all the ways gender bias seeps into both our algorithms
and society, how do we create a world that is more equal? *What
Works* uses behavioural science to map out the ways we can start
dismantling bias through design, right now.

Who Can You Trust? by **Rachel Botsman:** This compelling
book investigates the role of trust in the contemporary digital
ecosystem. Focusing on the impact of the gig economy,
cryptocurrencies and automated vehicles, Rachel Botsman
investigates how the institutions and ideas we put our faith in are
adapting to and shaping our new world.

Why I'm No Longer Talking to White People About Race
by **Reni Eddo-Lodge:** This powerful book dives into structures
of race and racism in Britain, and the ways white people are
complacent in this system. Important for thinking about women
and tech, she discusses how advances for white women often
exclude women of colour – how can we ensure we fight this kind
of white feminism in AI?

Women Invent the Future edited by **Doteveryone:** Martha Lane Fox's interview talks about the Doteveryone book and here it is. This rousing collection of sci-fi writings (six stories, one poem) creates a space for seven women to imagine our technological future. It discusses fertility, work, identity and love, but more than that it is about the possibilities – the victories and consequences – of what is to come.

You Look Like a Thing and I Love You by **Janelle C. Shane:** This book had me in stitches. Through her hilarious experiments, real-world examples and illuminating cartoons, Janelle C. Shane explains how AI understands our world, and what it gets wrong. More than just a working knowledge of AI, she hands readers the tools to be sceptical about claims of a smarter future.

Documentaries, Newsletters and Podcasts

***AI Now*'s yearly reports:** These spotlight the community groups, researchers and workers demanding a halt to risky and dangerous AI technologies and offer twelve recommendations on what policymakers, advocates and researchers can do to address the use of AI in ways that widen inequality.

AlphaGo (2017) by **Greg Kohs:** remember AlphaGo from Chapter Two? This documentary tells the whole story of the AI-powered victory and asks questions about what it means for the future of AI and our relationship with it.

Azeem Azhar's Exponential View: My favourite newsletter for stealing ideas to look smart at work or important dinner parties, Azeem carefully deconstructs the current state of play and shares complex but accessible commentary. His podcast is a who's-who of the industry and crucial listening if you'd like to feel like you are in the room with some of the authors I've picked out above.

CognitionX Newsletter: All the breaking news in the AI community delivered daily to your inbox. From the future of health to the impact on climate to business innovations, it's a great product to keep AI front of mind so you can stay on top of developments.

Deepmind: The Podcast: This is a super series from Deepmind, hosted by Hannah Fry, and will give you an insight into what the world's leading AI company is working on.

The Great Hack (2019) directed by **Karim Amer and Jehane Noujaim:** This documentary traces the Cambridge Analytica scandal, focusing particularly on the ways that new forms of data collection and analytics have unprecedented power to shape contemporary governance. Even for those who know the story, the documentary delves deep into the roles of particular figures at the company, and the ways the personal – data and beyond – is deeply political.

The Rise of the Machines: **Beth Singler** has produced four short documentaries on AI with the University of Cambridge and DragonLight Films, each one tackling a big question we have about AI. The first, *Pain in the Machine*, deals with whether robots could, and should feel pain. *Friend in the Machine* asks whether we could ever be real friends with robots. *Good in the Machine* is about AI ethics, and *Ghost in the Machine* ponders in just under fifteen minutes the really small question of whether AI could ever be conscious!

Fiction

The Foundation Trilogy by **Issac Asimov:** A monumental work in the history of science fiction, Asimov's Foundation trilogy is the epic tale of the fall and rise of future galaxies. Asimov is famous for his **Three Laws of Robotics**, which form the organising principle for all Asimov's robotic-based stories. Parts of his world and especially his laws, below, have since turned into fact, influencing developments in AI ever since:

1. A robot may not injure a human being or, through inaction, allow a human being to come to harm.

2. A robot must obey the orders given to it by human beings except where such orders would conflict with the First Law.

3. A robot must protect its own existence as long as such protection does not conflict with the First or Second Laws.

There is also a fourth law, worth noting because the machines themselves come up with their own law about protecting humanity.

Asimov's short story collection *I, Robot* contained elements that provided the inspiration for the pretty rip-roaring feature film of the same name starring Will Smith.

Frankisstein by **Jeanette Winterson:** Retooling *Frankenstein* for the current day, Jeanette weaves together poignant and funny threads to tell a story of gender, technology, love and innovation.

Hidden Figures directed by **Theodore Melfi:** Based loosely on the book by **Margot Lee Shetterly**, this movie tells the fascinating true story of three black women working for NASA during the Space Race – Katherine Johnson, Mary Jackson and Dorothy Vaughan (my hero, as you know already!). Navigating racism and sexism, each woman had to fight discrimination simply to do her job – a pattern still in force for black women today. I watch this on every long journey I take and sob.

The Hitchhiker's Guide to the Galaxy by **Douglas Adams:** One of the most famous sci-fi sagas of all time, *The Hitchhiker's Guide to the Galaxy* blends satire with true multi-galaxy pizzazz! Impossible to summarise, I implore you to delve into this weird and wonderful world for yourself.

The Power by **Naomi Alderman:** Weaving together the story of four people navigating a new society, where, overnight, girls become more powerful than men, *The Power* questions our ideas about strength, leadership, empowerment and, of course, girlhood. I felt it was a great thought experiment for what would happen if AI gave women equal or greater powers to men. I don't know if this is how Alderman intended readers to think but I couldn't help but draw parallels with this technology.

Mr Penumbra's 24-Hour Bookstore by **Robin Sloan:** In this innovative and uber-cool novel, Sloan has managed to create a world that simultaneously reveres print books while looking to the digital future. Set in silicon's world capital, San Francisco, carefully crafted characters explore what knowledge means for the modern world and who is granted access to it.

Murmur by **Will Eaves:** In Chapter Two, I touched on the work of Alan Turing, and *Murmur* is the award-winning genre-bending novel based loosely on his life. The book, like Turing's life, is a story of ideas, creativity, sexuality, pain, possibility and invention.

Trouble with Lichen by **John Wyndham:** This 1960s sci-fi novel imagines a world in which a young female scientist discovers a lichen that give women longer to live, and goes on to try to upend the traditional roles of women in her society. I loved reading this and replacing the lichen with AI and playing out how AI could be the ultimate emancipator, freeing women from the status quo.

Activities, Events and Courses

CogX Festival: Each June, the company I co-founded invites you to come and listen to the best thinkers in the world. In 2019, 20,000 visitors joined 600 speakers in London's Kings Cross to discuss the impact technology is having on industry, government and society. Join us at the next event.

Elements of AI: This course from the University of Helsinki and Reaktor is a fantastic place to start learning more about the practicalities of what AI is and what it isn't. You'll learn a bit more about the basic maths behind the technology, including simple introductions to things like neural networks and classifiers. No programming or maths background required, and it's free, so you can give it a go without worrying.

Founders4Schools: If you are in the UK and you'd like to get some AI experts to come and speak at your school, this charity is a great way to find and invite leaders to visit and talk about their experiences at no cost.

Machine learning courses from Andrew Ng on Coursera: There are two courses. One for business people, one that's a lot more technical. Millions have used these courses to understand more about machine learning, and if you're considering implementing, or even building AI in the future, then this is the best place to start.

Machinelearningforkids.co.uk: Don't be put off by its name saying it's 'for kids'. I genuinely think it's one of the most useful exercises I did when trying to understand AI. This free tool will help you understand how to train the computer to recognise text or images and even make it into a game.

AI Master's Programmes and Conversion Programmes: You can apply for these courses at twenty-eight universities from any degree, not just STEM courses. Instigated by Dame Wendy Hall in order to encourage more people from diverse backgrounds into AI, this is the perfect course for you if you did a humanities subject and want to learn about AI. There are 1,000 means-tested scholarships available per year, prioritised for black, female and disabled students. You can find out more on the Office for Students Website and search for **#JoinyourAIFuture**.

S2DS and Faculty.ai: These are two of Europe's largest data science training programmes. They both take exceptional analytical minds who have already completed STEM PhDs and MScs and helps create Data Science and AI experts.

Tech Mums: These courses, founded by **Dr Sue Black**, have been specially created for self-confessed technophobe mothers to take the mystery out of technology. Whether it's helping you to reconnect with old friends via social media, chatting to your child about online safety, or finding out how to use technology to help you at work, two hours a week means that they can guide you through a huge range of technological possibilities. And it's completely free!

Teens in AI: Are you in the UK, under eighteen years old and ready to get your hands dirty? Join one of Teens in AI's own hackathons, accelerators and bootcamps, or just become part of their gang and benefit from mentoring, talks, company tours, workshops and networking opportunities.

ACKNOWLEDGEMENTS

I write this with one hand, very slowly, so as not to move my arm too much as it will wake my eighteen-day-old son. Otis arrived three weeks earlier than expected, during the time I'd naively set aside to finish this book. I'm happy to say he has clearly always been in charge! Life with Otis is magical and unpredictable. I have a new centre around which to revolve, and my heart has well and truly exploded. I couldn't be more thankful for him, and to all the women in my life that made it seem even possible to contemplate having a baby.

Thank goodness I realised right from the start that writing this book was never going to be a solo act. In the history chapter, I wrote about how projects are almost always the result of teamwork and awesome women coming together. This book is no exception.

Without Michelle Kane, there wouldn't be a book. I know everyone says that about their editor, but in my case it's genuinely true – I'd never even contemplated writing more than emails before we met. Michelle was in that audience at the Riposte event I told you about in the foreword. She heard my message loud and clear and wanted to make sure other women did too. She gave me the platform to write this book for you. Thanks to my agent, Karolina Sutton at Curtis Brown, I have become an author guided by the most experienced hands in the industry.

Enter my heroes: Gemma Reeves and Eve Kraicer.

Gemma, my co-writer, a teacher and a novelist with an enviable skill for storytelling and breaking down difficult subject matter. Gemma is a self-confessed technophobe and this is part of what made her such a perfect person to veto, or validate, the stories I wanted to tell. Her ability to turn my upside-down thoughts and wonky word constructions into flowing sentences made this book what it is today.

Eve, our researcher, had just finished her dissertation in Gender, Media and Culture, when we met. I loved that she was writing about AI in relation to the #MeToo movement, and knew she'd bring the kind of curiosity and scrupulous attention to detail necessary to make this book sing.

Jess Wade is who you want on your side if you are embarking on writing a book. Honest but encouraging in equal measures is rare but being a physicist and illustrator maybe rarer – Jess is who we have to thank for bringing to life the interviews with her exquisite illustrations.

It's important to acknowledge the part Charlie Muirhead and our CognitionX investors including chair Joanna Shields played in my being able to navigate the AI landscape and bring these insights to you. I'm incredibly grateful for their support and understanding.

I didn't know how formative meeting Sophie Smith at thirteen years of age would be: from making me realise I could get an A* at maths if I studied my own way rather than how the teacher prescribed, to supporting my recent journey as I tried to understand the political repercussions of my own thought patterns, she has been a tremendous influence and much-needed support.

Without feedback from Hazel Falck, Lucy Luscombe, Issy Ferrandez, Emily Chappell, Jonnie Penn, Cheryl Clements, Claire Craig, Catherine Breslin, Verity Harding, Radhika Iyer, Zoe Berman, Eleonora Harwich, Priya Lakhani, Emma Sinclair, Dame Wendy Hall, Sana Khareghani, Gina Neff, Azeem Azhar, Scarlett Curtis and crucially Jane Procter my exceptional mother this book would be a 2D version of itself.

And lastly, Ed, without whom I'd be a 2D version of myself. Before I met him, I used to listen to a Marina and the Diamonds song every morning in the shower. I'd belt out the lyrics, 'I am not a robot/ Guess what, I'm not a robot.' But I think in many ways that I was a robot, desperately trying not to be. I was working every hour and ignoring everything else. Then I met Ed, the father of our son Otis, and my world changed. So although I wrote a book called *How to Talk to Robots*, it's Ed I have to thank for everything, as it's because of him that I live like a real human being, not a robot.

NOTES